MEDIA AND COMMUNICATIONS - TECHNOLOGIES, POLICIES AND CHALLENGES

BRIDGING THE DIGITAL DIVIDE IN INDIAN COUNTRY

FEDERAL EFFORTS

MEDIA AND COMMUNICATIONS - TECHNOLOGIES, POLICIES AND CHALLENGES

Additional books in this series can be found on Nova's website under the Series tab.

Additional e-books in this series can be found on Nova's website under the e-book tab.

MEDIA AND COMMUNICATIONS - TECHNOLOGIES, POLICIES AND CHALLENGES

BRIDGING THE DIGITAL DIVIDE IN INDIAN COUNTRY

FEDERAL EFFORTS

REID LOUTON
AND
JUDE SULLIVAN
EDITORS

nova publishers
New York

For permission to use material from this book please contact us:
Telephone 631-231-7269; Fax 631-231-8175
Web Site: http://www.novapublishers.com

Additional color graphics may be available in the e-book version of this book.

Library of Congress Cataloging-in-Publication Data

ISBN: 978-1-62948-170-8

Published by Nova Science Publishers, Inc. † New York

CONTENTS

PREFACE

The lack of communications services in Indian Country – be it high speed internet or "broadband", traditional wireline phone service, mobile service, radio broadcast, or TV broadcast service – is well known. As the Federal Communications Commission (FCC) has observed previously, "by virtually any measure, communities on Tribal lands have historically had less access to telecommunications services than any other segment of the population." The lack of robust communications services presents serious impediments to Tribal Nations' efforts to preserve their cultures and build their internal structures for self-governance, economic opportunity, health, education, public safety, and welfare. This book examines federal efforts to bridge the digital divide in Indian country with a focus on the internet infrastructure in native communities and equal access to e-commerce, jobs and the global marketplace.

Chapter 1 - The Office of Native Affairs and Policy's 2012 Annual Report is intended to provide the Commission with a review of the unprecedented level of coordination, engagement, and training with Tribal Nations and inter-Tribal government associations that took place in 2012. Because this is the first such report that the Office of Native Affairs and Policy (ONAP) has produced in its first two and one-half years of existence, the Annual Report also provides sufficient background and historical perspective to fully appreciate the magnitude of the Commission's commitment to Indian Country.

The Annual Report illustrates the Commission's efforts to address the digital divide in Indian Country, the progress that the Commission has made, and why that progress requires even more coordinated efforts in the years to come. ONAP's responsibilities encompass the consultation efforts of the entire agency, and the involvement and contributions of staff and managers of the Bureaus and Offices across the Commission will be clear throughout this

Annual Report. This ongoing partnership and commitment across the Commission to work with Tribal Nations is unprecedented in many ways.

The Commission shares a government-to-government trust relationship with Tribal Nations, which guides both the character and quality of the work that ONAP and the rest of the agency performs with Tribal Nations. The Commission has quickly earned a reputation among the Tribes as being one of the leading institutions of the federal government when it comes to creativity and determination to solve the persistent and pervasive problems related to the lack of 21st century technologies on Tribal lands. On several occasions in the 13 years that the Commission has focused its efforts to work with Tribal Nations, particularly since the creation of ONAP, many Tribal leaders have stated that the communications challenges on Tribal lands are often found in many different Tribal regions of the nation, but that in every instance the solutions to those problems are quite unique and very contextual, even on a regional basis. “One size fits none” is a constant refrain in the nation-to-nation work of ONAP.

The Annual Report will also illustrate that, in many respects, while many efforts to address the digital divide in Indian Country are in major motion, there is much more still to be done, new initiatives to undertake, and future milestones to achieve. Communications technologies and modern media platforms, such as broadband, hold the potential to level many of the negative impacts that history has visited on Tribal Nations. The Commission, through ONAP and ONAP’s coordination with the other Bureaus and Offices, has laid the groundwork for a path forward, and this Annual Report illustrates why future progress is both needed and justified.

Chapter 2 - This is an edited, reformatted and augmented version of a statement of Geoffrey Blackwell, Chief, Office of Native Affairs and Policy, Federal Communications Commission, presented October 6, 2011 before the Senate Committee on Indian Affairs.

Chapter 3 - This is an edited, reformatted and augmented version of a statement of Geoffrey Blackwell, Chief, Office of Native Affairs and Policy, Federal Communications Commission, presented June 8, 2012 before the House Committee on Natural Resources, Subcommittee on Indian and Alaska Native Affairs.

Chapter 4 - The “digital divide” is a term that has been used to characterize a gap between “information haves and have-nots,” or in other words, between those Americans who use or have access to telecommunications and information technologies and those who do not. One important subset of the digital divide debate concerns high-speed Internet access and advanced telecommunications services, also known as broadband. Broadband is provided by a series of technologies (e.g., cable, telephone wire, fiber, satellite, wireless) that give users the ability to send and receive data at volumes and speeds far greater than traditional "dial-up" Internet access over telephone lines.

Broadband technologies are currently being deployed primarily by the private sector throughout the United States. While the numbers of new broadband subscribers continue to grow, studies and data suggest that the rate of broadband deployment in urban/suburban and high income areas is outpacing deployment in rural and low-income areas. Some policymakers, believing that disparities in broadband access across American society could have adverse economic and social consequences on those left behind, assert that the federal government should play a more active role to avoid a "digital divide" in broadband access.

With the conclusion of the grant and loan awards established by the American Recovery and Reinvestment Act of 2009 (P.L. 111-5), there remain two ongoing federal vehicles which direct federal money to fund broadband infrastructure: the broadband and telecommunications programs at the Rural Utilities Service (RUS) of the U.S. Department of Agriculture and the Universal Service Fund (USF) programs under the Federal Communications Commission (FCC). Although the USF's High Cost Program does not explicitly fund broadband infrastructure, subsidies are used, in many cases, to upgrade existing telephone networks so that they are capable of delivering high-speed services. Additionally, subsidies provided by USF's Schools and Libraries Program and Rural Health Care Program are used for a variety of telecommunications services, including broadband access. Currently the USF is undergoing a major transition to the Connect America Fund, which is targeted to the deployment, adoption, and utilization of both fixed and mobile broadband.

To the extent that the 113th Congress may consider various options for further encouraging broadband deployment and adoption, a key issue is how to strike a balance between providing federal assistance for unserved and underserved areas where the private sector may not be providing acceptable levels of broadband service, while at the same time minimizing any deleterious effects that government intervention in the marketplace may have on competition and private sector investment.

In: Bridging the Digital Divide in Indian Country ISBN: 978-1-62948-170-8
Editors: R. Louton and J. Sullivan

Chapter 1

FEDERAL COMMUNICATIONS COMMISSION OFFICE OF NATIVE AFFAIRS AND POLICY 2012 ANNUAL REPORT*

Office of Native Affairs and Policy

INTRODUCTION AND OVERVIEW

The Office of Native Affairs and Policy's 2012 Annual Report is intended to provide the Commission with a review of the unprecedented level of coordination, engagement, and training with Tribal Nations[1] and inter-Tribal government associations that took place in 2012. Because this is the first such report that the Office of Native Affairs and Policy (ONAP) has produced in its first two and one-half years of existence, the Annual Report also provides sufficient background and historical perspective to fully appreciate the magnitude of the Commission's commitment to Indian Country.

The Annual Report illustrates the Commission's efforts to address the digital divide in Indian Country, the progress that the Commission has made, and why that progress requires even more coordinated efforts in the years to come. ONAP's responsibilities encompass the consultation efforts of the entire agency, and the involvement and contributions of staff and managers of the Bureaus and Offices across the Commission will be clear throughout this

---

* This is an edited, reformatted and augmented version of Office of Native Affairs and Policy, dated March 2013.

Annual Report. This ongoing partnership and commitment across the Commission to work with Tribal Nations is unprecedented in many ways.

The Commission shares a government-to-government trust relationship with Tribal Nations, which guides both the character and quality of the work that ONAP and the rest of the agency performs with Tribal Nations. The Commission has quickly earned a reputation among the Tribes as being one of the leading institutions of the federal government when it comes to creativity and determination to solve the persistent and pervasive problems related to the lack of 21st century technologies on Tribal lands. On several occasions in the 13 years that the Commission has focused its efforts to work with Tribal Nations, particularly since the creation of ONAP, many Tribal leaders have stated that the communications challenges on Tribal lands are often found in many different Tribal regions of the nation, but that in every instance the solutions to those problems are quite unique and very contextual, even on a regional basis. "One size fits none" is a constant refrain in the nation-to-nation work of ONAP.

The Annual Report will also illustrate that, in many respects, while many efforts to address the digital divide in Indian Country are in major motion, there is much more still to be done, new initiatives to undertake, and future milestones to achieve. Communications technologies and modern media platforms, such as broadband, hold the potential to level many of the negative impacts that history has visited on Tribal Nations. The Commission, through ONAP and ONAP's coordination with the other Bureaus and Offices, has laid the groundwork for a path forward, and this Annual Report illustrates why future progress is both needed and justified.

Top Tribal Consultation and Coordination Priorities for 2013

ONAP anticipates that 2013 will be as busy as 2012 with respect to Tribal policy initiatives, consultations, and trainings. ONAP has established its top consultation priorities for 2013, which include, in no particular order: (1) the Tribal Mobility Fund and support for wireless carriers seeking to serve Tribal lands; (2) the Tribal government engagement obligation under the new high-cost and Connect America Fund regulations; (3) the *Spectrum Over Tribal Lands Notice of Proposed Rulemaking* and proposals aimed at providing more efficient spectrum licensing on Tribal lands, including a Tribal Priority; (4)

cultural preservation and environmental review of non-compliant towers nationwide; (5) review of eligible telecommunications carrier (ETC) designation petitions filed by carriers seeking to provide service on Tribal lands; and (6) training and coordination of the various radio broadcast licensing opportunities now available to Tribal Nations under the Commission's Tribal Priority rules.

Tribal Mobility Fund

In 2013, the Commission expects to hold a reverse auction for Phase I of the Tribal Mobility Fund, which will distribute $50 million in one-time support for mobile service providers serving Tribal lands lacking 3G or 4G service.[2] To ensure broad and successful participation in the auction, ONAP and the Wireless Telecommunications Bureau (WTB) will partner to provide training for Tribal Nations on topics such as eligibility, opportunities for joint ventures, and the mechanics of the actual auction process.

Tribal Government Engagement Obligation

In 2013, ETCs will report for the first time on their compliance with the Tribal government engagement obligation adopted in the context of universal service reform.[3] This is an annual obligation that is still in its earliest stages. Supported communications providers are to meaningfully engage with the governments of the Tribal Nations on whose lands they serve. Pursuant to the Commission's order creating the obligation, ONAP is committed to developing best practices that examine, through potential efficiencies and regional commonalities, how best to bring industry and Tribal decision makers together to discuss critical deployment issues. In the context of training seminars as well as stand-alone meetings, ONAP will host and facilitate engagement meetings between ETCs and Tribal Nations.

Spectrum Over Tribal Lands Notice of Proposed Rulemaking

In March 2011, the FCC began a rulemaking proceeding designed to improve Tribal access to spectrum and to promote greater utilization of spectrum over Tribal lands through a number of proposals designed to expand

and enhance fixed and mobile wireless services on Tribal lands. The Commission sought comment on processes to provide Tribes with new opportunities to gain access to spectrum, including the establishment of a Tribal Priority (similar to the one in effect for broadcast radio), a formal negotiation process for secondary markets agreements, a build-or-divest process, and construction safe harbor provisions. Because access to spectrum continues to be a critically important need of Indian Country in closing the digital divide, ONAP, in coordination with WTB, will structure and carry out consultations related to these important regulatory proposals. As access to spectrum is a necessary predicate to being involved in the Mobility Fund reverse auctions, this is a top priority on multiple counts. Tribal Nations and major inter-Tribal associations, such as the National Congress of American Indians and the Southern California Tribal Chairmen's Association, remain supportive of the Commission taking action to implement the Tribal Priority provisions, and have offered to assist in further consultations on the formal negotiations, build-or-divest, and safe harbor proposed rules.

Cultural Preservation Review of Non-Compliant Towers

In 2013, ONAP and members of the intra-agency NEPA team[4] will partner to host and initiate consultations with Tribal Nations and inter-Tribal government associations regarding options and strategies for analyzing and addressing the status of various classes of towers that never went through historic preservation review under Section 106 of the National Historic Preservation Act. These towers were never reviewed by Tribal Historic Preservation Officers or Cultural Preservation Officials for potential adverse impacts on Tribal sites of religious and cultural significance, or Tribal "sacred sites." In dealing with "twilight towers" and other non-compliant towers, these consultations will involve issues and information of an extremely sensitive nature. Depending on the course chosen, these consultations will build upon a number of important successful Commission rules and policies[5], as well as Commission systems[6], that streamline and facilitate the review of tower sitings to minimize potential negative impacts on Tribal cultural and religious assets. It is important to stress here that this area of Tribal consultation is at its earliest stages of analysis and development. As the Commission begins to formulate its values, priorities, and positions with respect to towers that are non-compliant under Commission rules, it will do so by also seeking to understand the values, priorities, and positions of Tribal Nations.

Eligible Telecommunications Carrier Designation Petitions to Serve Tribal Lands

There are ten Tribally owned ETCs today and, as a result of recent policies adopted as part of universal service reform, there are new opportunities available for Tribes that may want to consider self-provisioning telecommunications and broadband services to their lands. For example, in the context of the Mobility Fund and the Tribal Mobility Fund, a Tribally owned applicant need only have its ETC designation petition filed at the short form application stage in order to participate in the auction. Opportunities such as these, as well as the ongoing challenges associated with the provision of telecommunications and broadband services on Tribal lands, has sparked a renewed interest in the opportunities available through ETC designation. There are currently two ETC designation petitions pending from Tribally owned or affiliated entities, and ONAP anticipates that the number of such petitions will likely increase. The requirements and responsibilities associated with ETC status, as well as the designation process, therefore, will be among ONAP's top consultation priorities for 2013.

Radio Broadcast Tribal Priority

With the adoption of the *3rd Report and Order* and the extension of the Commission's Tribal Priority licensing procedures in radio to full power commercial FM licensing, opportunity now abounds for Tribal Nations in virtually all areas of this important backbone and well-adopted communications technology. The Tribal Priority gives priority to Tribal Nations or Tribally-owned entities when proposing FM allotments and when filing AM and noncommercial educational (NCE) FM filing window applications.[7] Tribes now broadcast to their own communities with low power, full power, non-commercial educational, and commercial radio broadcast licenses. New radio licensing provides for Tribal Nations themselves to preserve and advance their languages and cultural values through a communications medium, to increase economic development opportunities through advertising and promotion, to provide job training and employment opportunities, and to promote the further adoption of new communications technologies, such as broadband. Through audio content development and on-line broadcasting of content previously available only over-the-air, Tribal Nations can also broadcast to their members living beyond their broadcast radius, such as their soldiers deployed abroad or other on-line listeners throughout the nation.

BACKGROUND

Description of the Problem

The lack of communications services in Indian Country – be it high speed internet or "broadband", traditional wireline phone service, mobile service, radio broadcast, or TV broadcast service – is well-known. As the Commission has observed previously, "by virtually any measure, communities on Tribal lands have historically had less access to telecommunications services than any other segment of the population."[8] The lack of robust communications services presents serious impediments to Tribal Nations' efforts to preserve their cultures and build their internal structures for self-governance, economic opportunity, health, education, public safety, and welfare.[9]

The sheer remoteness of regions in Indian Country can itself serve as a barrier to deployment of telecommunications services. Often, Tribal members have to drive to a nearby hilltop on a Reservation just to find a signal for mobile wireless service. And even after that, access may not be guaranteed.

> At the field hearings, the Commission heard first hand that the appalling lack of communications services on Tribal lands could even result in loss of life. For example, the Governor of the Pueblo of Jemez told of an incident in 1997 when a Tribal member, a young woman, on the Reservation had a seizure and did not have a telephone to call for help. Her boyfriend went to several neighboring homes to find a phone and, when he finally found a home with a phone, it did not work. He then ran to the sheriff's house, thinking that the sheriff could use his vehicle's radio to call in the emergency, but even the sheriff had to drive to higher ground to use his radio to contact Bureau of Indian Affairs (BIA) law enforcement – far away in Albuquerque. By then, far too many precious moments had been lost and it was too late; the young woman died.
>
> Summary of 1999 field hearing testimony.

Understanding the complexity of the digital divide in Indian Country requires an appreciation of the unique challenges facing Tribal Nations, which include deployment, adoption, affordability, and access to spectrum, as well as lack of investment dollars and access to credit and start-up or gap financing. Barriers to the deployment of communications services include rural, remote, rugged terrain, areas that are not connected to a road system, and difficulty in obtaining rights-of-way to deploy infrastructure across some Tribal lands – all of which increase the cost of installing, maintaining, and upgrading infrastructure. Affordability of communications services is affected by often endemic levels of poverty. Because Tribal Nations cannot easily collateralize assets that are held in trust by the federal government, and cannot easily access investment dollars, the ability to obtain credit and financing is limited. Despite these barriers, however, where Tribal Nations and their community members do have access to broadband, studies indicate that their rates of adoption and use are on par with, if not higher than, national averages.[10] And that fact is what breathes life into the Commission's Tribal policy agenda.

The Commission's Tribal Agenda before the Office of Native Affairs and Policy

The Commission's Tribal agenda was forged first during the term of Chairman William E. Kennard as part of efforts to close the telecommunications divide between Indian Country and the rest of the United

States. The Commission, through two major field hearings in 1999, learned about the lack of services in Indian Country and the impact this had on peoples' lives. Testimony provided in hearings, held in Albuquerque, New Mexico and Chandler, Arizona, painted a picture of communications services so lacking and so expensive to provide that the Commission ultimately took a number of critical regulatory actions.

The Commission heard testimony from a Navajo Nation Tribal Council Delegate about a disabled person who requested telephone service only to have the local phone company tell him that it would cost $15,000 to extend a line to his home on the Navajo Reservation. The Delegate estimated that it cost a minimum of $5,000 to connect a new telephone subscriber on the Reservation, underscoring the critical need for affordable basic telephone service.[11]

At the hearings, the Commission heard repeatedly how public safety, emergency services, medical care, education, and economic development lagged far behind the rest of the country because of the lack of communications services. It heard how employers were unwilling to locate on Reservations because of the absence of telecommunications infrastructure, and how it was difficult for Tribal enterprises to succeed. The Commission heard recommendations from presenters calling for the adoption of a policy statement recognizing the sovereignty of federally-recognized Tribes and establishing government-to-government relationships in order to bring improvements in communications services to Indian Country. The Commission also heard pleas for funding to counter the reluctance of carriers to provide services to rural and remote areas in Indian Country.[12]

Policy Initiatives

The hearings laid the foundation for a range of Commission policy initiatives to address the lack of communications services on Tribal lands. First and most importantly, in 2000, in response to the call for dealing with Tribes as sovereign nations and creating a framework for developing relationships with Tribal Nations, the Commission adopted a policy statement on establishing a government-to-government relationship with Indian Tribes. The *Tribal Policy Statement* recognized Tribal sovereignty, federal trust principles, and the importance of Commission consultation with federally-recognized Tribes. It also acknowledged the principles of Tribal self-governance and recognized "the rights of Tribal governments to set their own communications priorities and goals for the welfare of their membership."[13] The Commission's *Tribal Policy Statement* and the framework of its enumerated goals and principles guides the Commission's work with Tribal Nations to this day in consultation and coordination efforts exercising the Commission's government-to-government relationship with Tribal Nations.

> Upon adoption of the Tribal Policy Statement in 2000, Chairman Kennard said, "I believe that the most important thing that we have done is to adopt a policy statement that embraces tribal sovereignty, the federal trust responsibility and Indian self-governance. . . . We at the FCC promise to honor your fundamental right to self-governance, and we adopted this statement to ensure that this right is always respected and never infringed upon by the Commission." From Chairman Kennard's ITTI 2000 speech in St. Paul, MN.

Just one week after release of the *Tribal Policy Statement*, the Commission took its first steps to address the affordability of telephone service and what was then a 47 percent telephone penetration rate on Tribal lands. In the *Twelfth Report and Order*, the Commission acknowledged that existing universal service programs were not adequate to sustain telephone subscribership on Tribal lands. With the goal of providing basic telephone service for $1 a month on Tribal lands, the Commission created the enhanced universal service Lifeline and Link Up programs, today known as Tribal Lands Lifeline and Link Up. The Commission built upon the goal of the existing Lifeline and Link Up programs to help ensure that low-income consumers have and maintain access to basic telephone service by creating additional discounts available only for low-income consumers residing on Tribal lands. The Commission also adopted additional and more inclusive eligibility criteria to include income assistance programs in which low-income consumers living on Tribal lands would be more likely to participate, including BIA general assistance, Tribally-administered Temporary Assistance for Needy Families (Tribal TANF), and the National School Lunch Program's free lunch program.[14]

Also in the *Twelfth Report and Order*, the Commission made groundbreaking decisions about the ETC designation process that continue to this day to have a profound impact on the provision of telecommunications services on Tribal lands. ETC status is required for participation in the high-cost and low-income universal service programs and, once granted, provides access to those subsidy programs. Relying upon the unique federal trust relationship between the federal government and federally recognized Tribes, the Commission concluded that it may make the threshold determination of which entity – the Commission or the state – has jurisdiction to make ETC designations for providers serving on Tribal lands.[15] The process that the Commission created in 2000 has facilitated the designation of ETCs on Tribal lands, including the designation of Tribally owned ETCs, providing access to universal service funding for the provision of telecommunications service on Tribal lands.

In yet another action designed to increase the availability of telephone service on Tribal lands in 2000, the Commission adopted rules to provide incentives for wireless telecommunications carriers to serve consumers on Tribal lands. Specifically, the Commission established Tribal lands bidding credits that were available in auctions markets that contained qualifying Tribal areas with telephone penetration rates below 70 percent. That initial threshold level has since been raised to 85 percent. To qualify for the credit, winning bidders were required to use the license to deploy facilities and offer service to qualifying Tribal areas and to obtain Tribal consent to such deployment. [16]

Beginning in 2000, a major focus of the Commission was a review and update of its environmental and historic preservation processes for compliance with, among other legal requirements, Section 106 of the National Historic Preservation Act.[17] Section 106 is the authority by which federal agencies consult with Tribes to avoid and minimize impacts of federal undertakings upon sites of religious and cultural significance to Tribal Nations. The pervasive concern across Indian Country at that time was that the siting of communications towers, including mobile wireless and broadcast towers, were placing Tribal sacred sites at risk as they were not being properly reviewed. At the same time, the wireless and tower industries were seriously concerned that an onerous review of towers would impede the deployment of these critical services. With such controversial attention, Congressional oversight hearings, and potential litigation, the Commission undertook to address Tribal interests in a series of regulatory actions.

First, in 2001, the Commission adopted a Nationwide Collocation Agreement, streamlining the review processes for collocating on existing infrastructures.[18] Learning of the myriad Tribal concerns in the process of adopting this agreement, WTB and the Consumer and Governmental Affairs Bureau (CGB) coordinated in 2003 and 2004 on a consultation team with Tribal leaders and Tribal Historic Preservation Officers from throughout the nation. In late 2004, these efforts culminated with the adoption of a new Nationwide Programmatic Agreement (NPA) for streamlining the tower siting process.[19] The NPA, established between the Commission, the Advisory Council on Historic Preservation, and the National Council of State Historic

Preservation Officers, adopted new streamlined review protocols and underscored the Commission's commitment to see compliance with Section 106 in the protection of Tribal sacred sites.

Concurrent with the NPA, the Commission undertook two other actions as a result of the award-winning course of consultation and coordination with Tribal Nations on these important concerns. First, the Commission created the Tower Construction Notification System (TCNS), an online, double-blind, and private notification system that automatically creates a level of information flow between the Tribal Nations and tower constructors.[20] TCNS enables Tribes to notify tower constructors of proposed tower constructions that may require more review based on areas of Tribal concern, providing the capability for the further exchange of information.

On February 5, 2005, the Commission also entered into a Best Practices Agreement with the United South and Eastern Tribes, Inc. (USET), an inter-Tribal government association of the Tribal Nations from Maine to Florida to Texas.[21] USET, along with the National Congress of American Indians, was the leading organization in fostering the consultation with the Commission on the Section 106 concerns. A year earlier, the Commission and USET had entered into a Memorandum of Understanding on development of certain principles, procedures and best practices.[22] The NPA, TCNS, and USET Best Practices Agreement together provided expedited new processes for the review of towers for compliance with Section 106. These actions addressed the potential impact on Tribal sacred sites nationwide by the siting of communications towers in a comprehensive method, and engaged the Tribal Nations both prior to siting and in the course of siting – a process that, that many years later, has become a high standard and example to other agencies across the face of the federal government.

In 2008, the Commission adopted an emergency cap on payments to competitive ETCs under the universal service high-cost program that was intended to stem the growth of the Fund until the Commission adopted comprehensive reform. In the context of explosive growth of competitive ETCs accessing high-cost program support, the Commission had found that such a cap was necessary to ensure the future viability of the Universal Service Fund. Noting both the extremely low telephone penetration rate and the high level of poverty on many Tribal lands, however, the Commission waived the cap on Tribal lands under what was known as the Covered Lands Exception. The Commission recognized that higher levels of high-cost support were necessary to facilitate the expansion of wireless service on Tribal lands.[23]

Indian Telecom Training Initiative

The Commission also understood the importance of bringing together Tribal leaders, federal government officials, and industry to explore how communications services could be most effectively deployed in Indian Country. In September 2000, the Commission co-sponsored a seminal national conference in St. Paul, Minnesota, called the Indian Telecom Training Initiative 2000 (ITTI 2000). More than 600 people attended, including representatives from 135 federally recognized Tribes. More than 50 experts from federal government agencies, Tribal communities, the private sector, and foundations provided technical, financial, and regulatory information to inform Tribal leaders about the telecommunications industry. Chairman Kennard addressed the conference and articulated the Commission's commitment to Indian Country. As part of the Commission's effort to ensure that the voices of Tribal Nations were heard at every level of the Commission, Chairman Kennard announced that a new official position at the Commission, the FCC Liaison to Tribal Governments, would serve as the principal contact between Tribal Nations and the Commission.[24]

> "This conference is the culmination of the work that we have done at the FCC during my tenure as Chairman to close the telecommunications divide between Indian Country and the rest of America. Yet it is, in many ways, just the beginning of our work to make sure that no one living in Indian Country is left behind in the Information Age."
>
> "You must insist that your voices are heard. You must insist on an institutional commitment from the FCC. You must insist that you get as much attention as the armies of industry lobbyists there."
>
> "Together, we can make sure that the first Americans on this continent are not the last Americans to enjoy the wonders of the Internet."
>
> From Chairman Kennard's ITTI 2000 speech in St. Paul, MN.

In June 2001, the Commission co-sponsored a second ITTI conference in St. Paul, Minnesota – "Doing Business in Indian Country" – this one targeted to business and industry leaders in the areas of marketing, finance, sales, and business development in the telecommunications and utility industries. Speakers also included Tribal government leaders, Tribal elders, and Tribal college professors. The conference was developed to address industry's questions on initiating contact and developing programs with Tribal governments. The conference included sessions on Tribal priorities and business opportunities in telecommunications, how to work with Tribal governments, Tribal sovereignty issues, and community protocols and concerns as they relate to building telecommunications and utilities infrastructure and markets.

A third national ITTI conference, directed at Tribal leaders, was scheduled for October 2001. Due to the effects of the events of September 11, 2001, ITTI 2001 was initially postponed and ultimately cancelled. Many calls from every corner of Indian Country asked the Commission to continue and further develop its Tribal training initiatives.

Indian Telecommunications Initiatives

In 2002, under the leadership of Chairman Michael Powell, the Commission shifted its focus from large national meetings to smaller, regional events, re-named the Indian Telecommunications Initiatives (ITI) Regional Workshops and Roundtables. ITI's goals were threefold – to increase the telephone penetration rate on Tribal lands; to expand the infrastructure necessary to provide telecommunications services on Tribal lands; and to inform consumers in Indian Country about financial support available through federal government programs, including the universal service fund. ITI's intended outcomes included increasing Internet access, improving access to emergency and long-distance medical services, and enhancing education and employment opportunities for residents of Indian Country. As Chairman Powell said at the time, "The FCC remains committed to addressing telecommunications challenges facing Indian Country."[25]

The Commission conducted a number of activities as part of its ITI program, in coordination with Tribal governments, inter-Tribal organizations, industry, and others. But the cornerstone of ITI was a series of regional workshops and roundtables, which were held in the following locations:

Rapid City, SD: July 27-29, 2009
Salt Lake City, UT: July 16-17, 2008
Albuquerque, NM: July 10-11, 2007
Polson, MT: October 24-25, 2006
San Diego, CA: July 27-28, 2006
Albuquerque, NM: July 28-29, 2005
Coeur d'Alene, ID: November 9-10, 2004
Rapid City, SD: May 26-27, 2004
Reno, NV: July 17-18, 2003

In 2004, Chairman Michael K. Powell and United South and Eastern Tribes (USET) President Keller George signed a Memorandum of Understanding on formal Best Practices Agreement to be followed in siting communications towers and to cooperate in the establishment and maintenance of the Tower Construction Notification System. The FCC-USET best practices were created to promote cooperation between USET Tribes, the Commission, FCC Applicants, and entities subject to the jurisdiction of the Commission, and to protect Tribal properties of religious and cultural significance.

ITI focused on outreach, bringing together Tribal, federal government, and industry representatives to provide clear, practical, "how to" information about telecommunications services and infrastructure development that Tribes could use to gain access to critical telecommunications services. For example, the Commission distributed educational materials for Tribal consumers on issues such as the enhanced Tribal Lifeline and Link Up programs and how to file an informal complaint at the Commission. The Commission also distributed information about rules and policy initiatives affecting telecommunications services in Indian Country. ITI outreach activities also provided the Commission with an opportunity to establish beneficial relationships with Tribal governments, organizations and their members, and to listen and learn about the telecommunications needs of Tribal Nations.

Creation of the Office of Native Affairs and Policy

The National Broadband Plan signaled a new beginning for the Commission's Tribal policy work. Including dozens of recommendations for expanding the reach of broadband into Indian Country,[26] and incorporating the comments and input of Tribal leaders,[27] the National Broadband Plan breathed new life into the Commission's Tribal agenda. Broadly, the Plan recommended that the Commission should increase its commitment to government-to-government coordination with Tribal leaders and consider increasing Tribal representation in telecommunications planning. More specifically, the National Broadband Plan recommended the creation of a Tribal office within the Commission, whose role would include responsibility for fostering consultation with Tribal governments and leading the development and implementation of a Commission-wide Tribal agenda in coordination with the other Bureaus and Offices.[28]

Thus, the Office of Native Affairs and Policy (ONAP) was created by a unanimous vote of the Commission on July 29, 2010,[29] as the policymaking and physical embodiment of the Commission's commitment to Indian Country. The remaining months of 2010 saw the creation and the staffing of certain positions within ONAP. That summer and early fall, ONAP introduced the new Office to Tribal Nations by actually rolling it out in Indian Country, with managers and members of the newly formed staff traveling to meetings with Tribal leaders, a Tribally owned radio station, a Tribal utility authority, and an inter-Tribal association – the first of which was a meeting with the National Tribal Telecommunications Association, the association of Tribally owned and operated telecommunications companies.

In 2011, in close coordination with the Office of the Chairman and several other Bureaus and Offices, ONAP undertook to develop a renewed regulatory and policy agenda for Tribal Nations and Native Communities. On March 3, 2011, the Commission used its open public meeting agenda to hold Native Nations Day, which was groundbreaking in many ways. First, the Commission rolled out three major rulemakings affecting Tribal lands – the *Second Report and Order and Further Notice of Proposed Rulemaking* of the Tribal Priority in radio broadcasting, the launch of the *Spectrum Over Tribal Lands Notice of Proposed Rulemaking* for new wireless licensing opportunities for Tribal Nations and Tribal lands, and the launch of ONAP's omnibus *Notice of Inquiry* seeking comment on a wide range of issues related to communications services in Indian Country.[30] Second, for the very first time, elected Tribal leaders directly addressed the full Commission at an Open Meeting. A panel of four Tribal Leaders from across the nation illustrated the many issues of concern and the communications needs of Tribal Nations. Many more elected and appointed Tribal leaders were present in the audience. And third, Chairman Genachowski announced the members of the newly formed FCC-Native Nations Broadband Task Force. Native Nations Day was a rare sight in the Commission's Meeting Room and another seminal day in the history of the Commission's work with Tribal Nations.

Additional major rulemakings affecting Tribal lands were also released in 2011 and 2012.[31] Also in 2012, ONAP launched a renewed and strengthened outreach, training, consultation, and coordination with Tribal governments and inter-Tribal organizations. Within this report, much more detail will be provided on the Commission's new Tribal consultation and coordination programs.

> "A key component of consultation is to understand the challenges currently facing Tribal communities. We urge the FCC to visit Tribal lands and meet with Tribal leaders to see the needs of their communities firsthand. This will prove invaluable as the FCC partners with Tribes to develop and implement policy solutions. Effective consultation will depend upon the availability – and the ability – of the Office of Native Affairs and Policy to coordinate with Tribes to develop viable and genuine Tribal-centric solutions. Indian Country has been encouraged by the creation of the Office. However, to be credible and effective, it must be given sufficient authority over communications issues affecting Indian Country."
>
> From National Congress of American Indians President Jefferson Keel's statement on Native Nations Day 2011.

Thus, the creation of ONAP represented a major step forward in the Commission's own trust relationship with Tribal Nations. As the federal government shares a unique trust relationship with federally recognized Tribes,[32] this historic relationship requires the federal government to adhere to certain fiduciary standards in its treatment of and dealings with Tribal Nations.[33] In this regard, the federal government has a long-standing policy of promoting Tribal self-sufficiency and economic development, as embodied in various federal statutes.[34] As an independent agency of the federal government, the Commission formally recognizes its own general trust relationship with, and responsibility to, federally recognized Tribes.[35] The Commission also recognizes "the rights of Indian Tribal governments to set their own communications priorities and goals for the welfare of their membership."[36]

In a historic event, Tribal leaders addressed the full Commission for the very first time on Native Nations Day, March, 3, 2011. Seated (pictured left to right) at the presentation table were Lieutenant Governor Joe Manuel of the Gila River Indian Community, Chairman Robert Smith of the Pala Band of Mission Indians, Council Member Susie Allen of the Confederated Tribes of the Colville Reservation, and President Jefferson Keel of the National Congress of American Indians, who serves as the Lieutenant Governor of the Chickasaw Nation. Joining the panel was Geoffrey C. Blackwell, Chief of the Commission's Office of Native Affairs and Policy.

Therefore, the Commission has concluded that any approach to deploying communications services, removing barriers to entry, and increasing broadband availability and adoption must recognize Tribal sovereignty, autonomy, and independence, the unique status and needs of Tribal Nations and Native communities, the importance of consultation with Tribal government and community leaders, and the critical role of Tribal anchor institutions.[37]

The Office of Native Affairs and Policy has testified in four hearings on Capitol Hill:

April 5, 2011: "Closing the Digital Divide: Connecting Native Nations & Communities to the 21st Century," before the Committee on Communications, Science and Technology, U.S. Senate.

Oct. 6, 2011: "Internet Infrastructure in Native Communities: Equal Access to ECommerce, Jobs and the Global Marketplace," before the Committee on Indian Affairs, U.S. Senate.

December 1, 2011: "Deficit Reduction and Job Creation: Regulatory Reform in Indian Country," before the Committee on Indian Affairs, U.S. Senate.

June 8, 2012: "FCC's Rule on USF and Its Impact on American Indians & Alaska Natives," before the Subcommittee on Indian & Alaska Native Affairs, Committee on Natural Resources, U.S. House of Representatives.

The Mission of the Office of Native Affairs and Policy

In the Order creating ONAP, the Commission explicitly defined ONAP's role and responsibilities:

This office will be charged with bringing the benefits of a modern communications infrastructure to all Native communities by, among other things, ensuring robust government-to-government consultation with Federally-recognized Tribal governments and other Native organizations; working with Commissioners, Bureaus, and Offices, as well as with other government agencies and private organizations, to develop and implement policies for assisting Native communities; and ensuring that Native concerns and voices are considered in all relevant Commission proceedings and initiatives.[38]

Ensuring that Native voices are heard and taken into account at the Commission requires that ONAP conduct its mission in two places – on Tribal lands across the nation and in Bureaus and Offices across the Commission. Internally, ONAP's role is to coalesce Native input and coordinate with Bureaus and Offices, Commissioners' Offices, and the Chairman's Office, to address issues raised and the impacts on Tribal Nations and carriers serving Tribal lands. That is, ONAP helps facilitate, draft, analyze, and advise on policy issues affecting Indian Country at every level of drafting and decision making in the Commission. Externally, ONAP convenes and consults with Tribal leaders in Indian Country, meets with carriers and providers that serve Tribal lands, and brings their perspectives directly into the Commission and its policies. These roles are necessarily complimentary and inter-dependent – one cannot be accomplished without accomplishing the other. ONAP fuses both its internal and external roles by bringing Indian Country to the Commission and the Commission to Indian Country.

ONAP also fulfills a role for the Commission on Capitol Hill, as evidenced by ONAP's testimony at four Congressional hearings in two short years. In addition, ONAP helped to prepare Commissioner Clyburn for her testimony before the Senate Committee on Indian Affairs in 2012. While a broad range of issues affecting Indian Country was addressed, the primary focus of four of these five Congressional hearings was universal service reform and its impact on Tribal lands.

In short, ONAP has worked in partnership with the Bureaus and Offices of the Commission, and across and in partnership with Indian Country, to help create regulatory platforms for new and improved relationships and growth of opportunities between the communications industry and Tribes. ONAP has also worked to help create opportunities in the Commission's rules for Tribal Nations to own and operate communications entities and self-provision for, in many instances, self-provisioning is the only model that ensures such costly critical communications infrastructure and services will be delivered to Indian Country. A fundamental goal of the federal trust relationship is to provide opportunities to Tribal Nations and Native communities to exercise their own self-determination. And, as the Commission has previously stated, "Native Nations are intimately acquainted with their members' needs and have valuable insight into how to meet them."[39]

A common thread in ONAP's work, and therefore throughout this report, is teamwork. ONAP has focused since its inception on providing valuable and meaningful input and assistance. This means ONAP staff rolling up their sleeves, working side by side with colleagues in the other Bureaus and Offices, with the shared goal of ensuring that Native voices are heard and receive appropriate action in all relevant rulemakings and other proceedings at the Commission.

As the roots of the problems that led to the digital divide in Indian Country run deep and represent complexity involving the law, policy, and history, in many ways the work has just begun. An overview of these coordinated efforts appears below.

How the Commission Has Addressed the Needs of Indian Country Since the Inception of the Office of Native Affairs and Policy

A Renewed Regulatory Agenda

The Commission established ONAP to more fully realize the Commission's long-standing commitment to the trust relationship between the federal government and Tribal Nations. With ONAP helping lead the efforts, the Commission has worked ever more closely with Tribal governments and incorporated policy recommendations and comments from Tribal governments, Tribally-owned telecommunications companies, and regional and national inter-Tribal organizations. This commitment to the federal trust relationship is found in a number of specific policies and new opportunities for Tribal Nations, including reforms to the Commission's universal service programs and broadcast licensing rules. For example, the Commission adopted numerous Tribal-specific provisions in the Connect America Fund and the Lifeline and Link Up programs as part of its comprehensive universal service reform, and extended the established Tribal Priority rules to enable Tribal Nations to more easily obtain full power commercial FM radio licenses.

Native Nations NOI

One of the first policy initiatives which ONAP led was the launch of an omnibus Notice of Inquiry – just seven months after the Office was created – on a broad cross-section of issues that contribute to and result from the lack of robust communications services on Tribal lands. On Native Nations Day 2011, ONAP presented the Notice of Inquiry to the full Commission, with many Tribal leaders in attendance. In the Notice of Inquiry, the Commission for the first time articulated and sought comment on certain critical communications issues facing Indian Country, laying the groundwork to identify present and future solutions. These issues included greater broadband deployment, adoption issues, deployment issues, opportunities for Tribal self-provisioning of communications services, the need for a uniform definition of Tribal lands to be used across the Commission in rulemakings, and the importance of strengthening the Commission's government-to-government consultation process with Tribal Nations.[40] Many of these issues, and the comments received in response to the Notice of Inquiry, have informed subsequent Tribal policy initiatives. On many issues, ONAP continues to work with Tribal leaders, inter-Tribal organizations, and industry to develop the records for regulatory actions and policy initiatives.

Tribal Provisions in USF Reform

Inclusion of the Hawaiian Home Lands in the definition of Tribal lands

Adoption of a Tribal government engagement obligation.

Creation of the Tribal Mobility Fund, including a 25 percent bidding credit for Tribally owned or controlled ETCs and a special ETC designation provision.

Special consideration for remote areas of Alaska.

Limited exception to the phasedown of competitive ETC support for Standing Rock Telecommunications, Inc.

Priority for waiver petitions filed by ETCs serving Tribal lands.

Carrier reporting requirements to Tribal government.s

Questions in the FNPRM regarding the impact of proposed reforms on Tribally owned ETCs.

Adoption of a Tribal variable in the High Cost Loop Support regression analysis.

Adoption of an Alaska variable in the High Cost Loop Support regression analysis.

Universal Service Reform

In the context of the Connect America Fund and the reform of the universal service high-cost program, dozens of Tribal-specific questions were presented in the proposed rulemaking, and the Commission engaged Tribal governments and associations, Native institutions, and community leaders throughout the country. The Commission established an open and direct dialogue through ONAP and the various Bureaus and Offices across the agency. As a result, the reforms adopted by the Commission recognize the unique circumstances and challenges of providing telecommunications and broadband services on Tribal lands and take numerous steps to address these challenges.[41]

With the creation of the Mobility and Tribal Mobility Funds, the Commission established a universal service support mechanism dedicated expressly to mobile services. Phase I of the Mobility Fund offered approximately $300 million in onetime support to address gaps in mobile service availability by supporting the build-out of current and next-generation mobile networks in areas where these networks are unavailable. Phase I of the

Tribal Mobility Fund will offer an additional $50 million in one-time support targeted exclusively for advanced mobile services on Tribal lands. Phase II of the Mobility Fund will offer $500 million annually for ongoing support of mobile services, with up to $100 million of this amount designated annually and exclusively for support to Tribal lands. These set-asides for Tribal lands represent an extraordinary commitment by the Commission for the foreseeable future. Mobile providers serving Tribal lands are eligible for funds available under both the Mobility and Tribal Mobility Funds.[42] There exists great potential for Indian Country in both the Mobility and Tribal Mobility Funds.

Another important development to identify and address root causes behind the lack of services on Tribal lands is the adoption of a Tribal government engagement obligation for all ETCs either currently providing service or seeking to serve Tribal lands. This obligation now requires carriers to "meaningfully engage" with the Tribal governments on whose lands they serve, on a series of topical areas and issues critical to the deployment of communications services on Tribal lands. At a minimum, the *USF/ICC Transformation Order* stated that such discussions must include: (1) a needs assessment and deployment planning with a focus on Tribal community anchor institutions; (2) feasibility and sustainability planning; (3) marketing services in a culturally sensitive manner; (4) rights of way processes, land use permitting, facilities siting, environmental and cultural preservation review processes; and (5) compliance with Tribal business and licensing requirements.[43]

The Commission adopted this obligation in response to commenters that included the National Tribal Telecommunications Association, the National Congress of American Indians, the Affiliated Tribes of Northwest Indians, and Native Public Media, who emphasized the critical role that Tribal consultation and engagement play in the successful deployment of service on Tribal lands. ONAP coordinated numerous meetings with industry representatives both before and after the adoption of this engagement requirement, involving many communications companies and providers on Tribal lands, as well as many state communications industry associations, including Alaska, Montana, South Dakota, New Mexico, and Arizona. The Commission did not supplant its own ongoing obligation to consult with Tribes on a government-to-government basis, but instead recognized the important role that all parties play in expediting communications service to Tribal lands.[44] The Tribal government engagement obligation holds great promise and represents an opportunity for Tribal governments and ETCs to coordinate on many issues critical to the deployment and adoption of communications services on Tribal lands.

Finally, the adoption of a Tribal variable in the regression analysis used to calculate High Cost Loop Support recognized that there are additional costs associated with the provision of telecommunications services on Tribal lands. In the *USF/ICC Transformation Order*, the Commission adopted a benchmarking rule intended to moderate the expenses of rate-of-return carriers with very high costs compared to their similarly situated peers, while further encouraging other rate-of-return carriers to advance broadband deployment.[45] In its subsequent *Benchmarks Order*, the Wireline Competition Bureau built on the analysis provide in the *USF/ICC Transformation FNPRM*, but also included a number of changes in response to comments received and further analysis by the Bureau. One such change, incorporated as result of input received from some of the Tribally owned ETCs, was the adoption of a Tribal variable that takes into account the higher costs associated with providing service on Tribal lands.[46] The adoption of the Tribal variable, as well as the adoption of an Alaska variable, underscores the often transformative impact of ensuring that Tribal voices are heard and incorporated into rulemakings and policies affecting Indian Country.

Lifeline and Link Up

Since 2000, the Tribal Lands Lifeline and Link Up programs have provided invaluable assistance in helping to dramatically improve access to telephone service for low-income residents of Tribal lands, many of which face often endemic levels of cyclical poverty. In the context of its 2012 reform of these programs, the Commission recognized the unique circumstances facing Tribal lands and adopted a number of Tribal-specific provisions. For example, the Commission adopted the Food Distribution Program on Indian Reservations (FDPIR), commonly referred to as the "commodity program," as an eligibility criterion. The Commission learned that, because many Tribal elders opted for this program rather than food stamps, they had been excluded from receiving the benefits of Lifeline and Link Up support. As another example, the Commission preserved the Link Up program on Tribal lands while eliminating it elsewhere, acknowledging the significant telecommunications deployment and access challenges that persist today on Tribal lands. The Commission also adopted a broadband adoption pilot program, in an effort to gather data on broadband adoption and deployment among low-income consumers, and directed that at least one pilot application providing service on Tribal lands be accepted.[47] Ultimately, applications from two Tribally owned ETCs were accepted into the pilot program.[48]

On December 19, 2012, the Wireline Competition Bureau selected 14 participants for the Lifeline Broadband Adoption Pilot Program. Two Tribally owned ETCs were selected – Gila River Telecommunications, Inc. (GRTI), serving the Gila River Indian Community, and Hopi Telecommunications, Inc. (HTI), serving the Hopi Tribe.

Both Tribally owned carriers will gather data on broadband adoption and deployment among low-income consumers on Tribal lands. GRTI's project will study the effects of varying subsidy amounts and choices for broadband speed. HTI will study the effects of varying end-user charges based on broadband speed and access to discounted equipment.

Spectrum Over Tribal Lands

Access to spectrum, and the attendant access to wireless voice and data services so essential to economic development, education, and health care, is another critically important issue on Tribal lands. The Commission has an ongoing rulemaking proceeding designed to improve Tribal access to spectrum and to promote greater utilization of spectrum over Tribal lands.[49] In its *Spectrum Over Tribal Lands NPRM*, the Commission sought comment on a number of proposals designed to expand and enhance fixed and mobile wireless services on Tribal lands – all with the objective of promoting greater use of spectrum over Tribal lands.

In the *Spectrum Over Tribal Lands NPRM,* the Commission proposed a Tribal priority (similar in concept to the priority in effect for radio broadcast licensing) that would be available only for unserved or underserved Tribal lands for qualifying Tribal entities, designated as such by the relevant Tribal government. For such Tribal lands within a geographic area covered by an unassigned license, the Commission sought comment on a proposal that a Tribal priority would permit a qualifying Tribal entity to proceed in licensing without proceeding to competitive bidding. To address difficulties that Tribes have had in securing access to spectrum rights held by existing wireless licensees whose licenses cover Tribal lands, the Commission also sought comment on a Tribal proposal for the creation of a formal negotiation process through which a Tribe that had been refused good faith negotiations regarding a secondary markets transaction within a wireless licensee's geographic area of license could require the licensee to enter into such negotiations. The Commission also sought comment on whether and how best to implement a build-or-divest process when a wireless licensee has fulfilled its construction requirements, but Tribal lands within the area of license remain unserved or underserved. The Commission additionally proposed a Tribal lands safe harbor build out provision in which a licensee would be deemed to have met its construction obligations for its entire service area if it provides a specified level of service to Tribal lands.[50]

Tribal Radio

In the broadcast arena, the Commission has adopted policies to significantly expand the number of radio stations owned by Tribes broadcasting to Tribal lands. The purpose is to promote sovereign rights of Tribes by enabling them to provide vital radio services to their communities and set their own communications priorities and goals. Recent Commission initiatives include establishment of a Tribal Priority (2010),[51] adoption of a methodology for identifying a licensing area that comports with the goals of the Tribal Priority for a Tribal Nation that lacks a land base, or has a small or irregularly shaped land base for the purposes of a broadcast licensing contour (2011),[52] adoption of a threshold qualifications process (2011),[53] and adoption of certain new low power FM (LPFM) rules tailored to Tribal lands (2012).[54] Each of these initiatives is intended not only to provide radio service tailored to specific Tribal needs and cultures, but to increase ownership of such radio stations by Tribes and Tribally-owned entities.

In northern Wisconsin, ONAP visited radio station WOJB-FM 88.9 FM, which broadcasts the Anishinaabe language in 100,000 watts from the reservation of the Lac Courte Oreilles Band of Lake Superior Ojibwe. As with visits to other stations, ONAP spoke with General Manager Sidnee Kellar and staff about financial and other challenges of operating a Tribally-focused community radio station in a rural or remote location.

Adjudicatory and Certification Matters

In addition to policy initiatives, ONAP focuses its efforts on working in concert with the other Bureaus and Offices on a variety of adjudicatory and certification matters. For example, ONAP works in coordination with the Wireline Competition Bureau (WCB) and the Wireless Telecommunications Bureau (WTB) on ETC designation petitions from Tribally-owned entities and publicly or privately owned entities seeking to serve on Tribal lands. ONAP also works with WCB on multiple types of adjudicatory matters or complex inquiries including, for example, E-rate appeals and rural health care questions from program applicants on Tribal lands. ONAP also works with the Bureaus on a variety of adjudicatory matters involving Tribally-owned ETCs. While by no means an exhaustive list, this is intended to provide a sense of the array of work involved in ONAP's mission – most of which is initiated by virtue of ONAP's routinely welcome presence on the ground in Indian Country and throughout the halls at Commission headquarters.

A Renewed Commitment to Consultation, Outreach, and Training

Chairman Genachowski's appointment of elected and appointed leaders from across Tribal Nations and senior staff members and decision makers from across the Commission to the FCC-Native Nations Broadband Task Force on March 3, 2011, brought to life one of the recommendations of the National Broadband Plan.[55] With its mission of, among other things, eliciting input to ensure that Native concerns are considered in all Commission proceedings related to broadband and developing recommendations for promoting broadband deployment and adoption on Tribal lands, the Task Force has been an invaluable resource in rulemaking proceedings critical to Indian Country, including most prominently the reform of universal service. In many respects, the work of the Task Force has just begun, as we continue to work together on broadband issues facing Tribal lands and as we launch new initiatives, such as developing and executing a Commission consultation policy and coordinating with external entities, including other federal departments and agencies – all of which falls squarely within the mission of the Task Force.

With the creation of ONAP, the Commission breathed new life into its consultation and training seminar initiatives in Indian Country. ONAP met with or presented before approximately 200 Tribal Nations in fiscal year 2012. As part of these initiatives, the Commission hosted six separate broadband and telecommunications training and consultation seminars on Tribal lands and in remote regions of the country. These ongoing "communications 101" and dialogue events are held for one to three days in length, depending on priorities, and are designed for Tribal leaders and decision makers, Tribal telecommunications, IT, and strategic planning professionals, as well as industry representatives. In fiscal year 2012, these events were hosted by the Commission in Washington State, California, Oklahoma, Wisconsin, and two locations in Alaska. The majority of these events were held on Indian Reservations and Tribal lands. Each of these training and consultation seminars included presentations and the active participation of managers and staff from throughout the Commission, including WCB, WTB, the Media Bureau, and the Office of the Managing Director. A more detailed overview of these consultation and training seminar initiatives is provided in one of the case studies of this report.

At any given time, ONAP is also involved in a wide array of project-driven consultations with Tribal Nations – from licensing issues to universal service reform to tower siting issues and beyond. These consultations range from short-term projects to ongoing, long-term projects. For example, ONAP worked with the Yurok Tribe to facilitate the award of an experimental license by the Office of Engineering and Technology. This license permitted the Tribe, in partnership with a wireless provider, to bring mobile broadband services to previously unserved portions of its Reservation along the Klamath River in northern California. With respect to long-term, ongoing projects of focus, ONAP works closely with Tribal Nations that own and operate their own broadband, broadcast, or telephone providers. ONAP works with Tribal Nations and their providers both individually and through their collective associations, on a broad range of matters affecting these Tribally owned providers.

For example, ONAP has worked closely with the National Tribal Telecommunications Association (NTTA), an association comprised of the ten Tribally owned ETCs, since ONAP's earliest days. This work has taken the form of attending meetings in Indian Country, helping to facilitate meetings at the Commission, and ensuring that the often unique circumstances facing Tribally owned and operated communications providers and ETCs are memorialized in the form of written comments and ex partes. This valuable input is then analyzed in coordination with other Bureaus and Offices and the Commission in rulemakings and individual adjudicatory matters. NTTA has coordinated closely with the Commission, through ONAP, throughout the groundbreaking reform of the high-cost portion of the universal service fund.

ONAP has also coordinated closely with Tribal Nations that own or control radio broadcast stations. This coordination and consultation also occurs both at the individual and association levels. The best example of this is the coordination between ONAP, the Audio Division of the Media Bureau, and Native Public Media (NPM), to focus on the many complicated legal and practical questions found in the Tribal Priority radio broadcast proceedings. NPM is an association of Tribal radio stations nationwide, that also works closely with the National Congress of American Indians (NCAI). NCAI is the oldest and largest inter-Tribal government organization in the country and has a highly respected Telecommunications Subcommittee. ONAP has played an important outreach and training role at the annual and quarterly meetings of NTTA, NPM, and NCAI.

ONAP plays important Tribal consultation and training roles at regular meetings of inter-Tribal government organizations and Tribal communications industry association, including (pictured from left to right) the National Tribal Telecommunications Association, Native Public Media and the National Congress of American Indians.

ONAP is also actively involved in the important and sensitive environmental and cultural preservation review issues associated with the siting of communications towers nationwide, including mobile wireless and broadcast towers. For example, in 2011, representatives from ONAP and WTB attended a meeting of the Tribal Historic Preservation Officers from across the northern plains states of Montana, North Dakota, and South Dakota. This meeting was held in the Black Hills of South Dakota, a region of significant sacredness to many Tribal Nations. In 2012, ONAP and WTB also took part in the Tribal cultural preservation "To Bridge A Gap" Conference, held annually in Oklahoma. More of these regional face-to-face meetings will be necessary across Indian Country in the coming year because of the critical issues associated with historic preservation review of various classes of deficient or non-compliant towers on Tribal lands. ONAP also maintains an important ongoing role in these critical issues through its membership on the Commission's intra-agency NEPA team. These examples represent but a fraction of the project-driven consultations in which ONAP is involved, in partnership with colleagues across the Commission.

In 2012, ONAP conducted meaningful Tribal Nation consultation and industry coordination work in places never before visited by the Commission. Certain aspects of ONAP's visit to Alaska in the summer of 2012 are perfect examples. Over the course of 10 days, ONAP staff crisscrossed the state of Alaska, from arrival in the Native Village of Kotzebue to the Native Village of Noatak, the city of Nome, Anchorage, the Native Village of Ruby, and the city of Fairbanks. ONAP staff met and engaged with Tribal Councils and Native Corporations, telecommunications providers large and small, inter-Tribal organizations, remote telemedicine and regional health care facilities, and police chiefs and fire chiefs responsible for public safety in the most remote environments—all packed into a detailed and aggressive agenda that would and did challenge even the most travel-hardy policy makers. There were sunset-less days and extreme weather, which included the largest single day of rainfall in the history of Nome, Alaska. ONAP hosted a town hall meeting in Nome attended by the general public which, in spite of the torrential rainfall and cold temperatures, was very well-attended. In the Native Village of Ruby, on a bluff above the Yukon River, ONAP met with the Ruby Village Tribal Council in the old traditional log home Council Chambers, in a meeting that was attended by practically every adult member of the community.

ONAP staff met many residents of the Native Village of Ruby, Alaska, as they gathered at the Community Hall for a meeting on telecom issues. Nestled above the Yukon River, Ruby is in the vast interior of Alaska. While not all the buildings have indoor plumbing, including the Community Hall, the hospitality of the Native Village of Ruby is legendary. The packed meeting covered many areas of the FCC's regulations and Tribal prerogatives.

The meeting resonated with all too familiar deployment and adoption issues. ONAP hosted two separate training seminars for Alaska Native Village representatives and communications providers alike – one in Anchorage, with a keynote address by U.S. Senator Mark Begich; and another in Fairbanks, hosted in cooperation with the Tanana Chiefs Conference, an inter-Tribal association of Alaska Native Villages.

ONAP met with the Tribal Council of the Village of Noatak on several telecom needs. The village is located nearly 100 miles north of the Arctic Circle in far northwest Alaska, and is accessible only by small aircraft which lands on a gravel runway that is so common across Alaska. Food, fuel, virtually everything is flown in. Low water levels make Noatak even inaccessible via boat; however the community spirit was warmly welcoming and engaging.

During one of the most extraordinary experiences of an office well familiar with the extraordinary, ONAP met with the Tribal Council of the Village of Noatak, 100 miles north of the Arctic Circle. Noatak is accessible only by air, as there are no roads beyond the edge of the small town and the river levels do not allow access by boat. Everything comes in by air, and the gas and grocery prices reflect it. ONAP and the other guests visited Noatak Village leaders along with representatives from Senator Mark Begich's office, the Alaska Telephone Association, GCI, and OTZ Telephone Cooperative, Inc., the region's local carrier. ONAP saw the daily operations of Noatak's local telemedicine facility and visited local telecommunications satellite and network facilities. ONAP staff witnessed first-hand the vast benefits that telemedicine brings to remote Alaska villages, where the ability of the Tribal health clinic to connect to physicians via broadband connections is literally the difference between life and death to Tribal village residents.

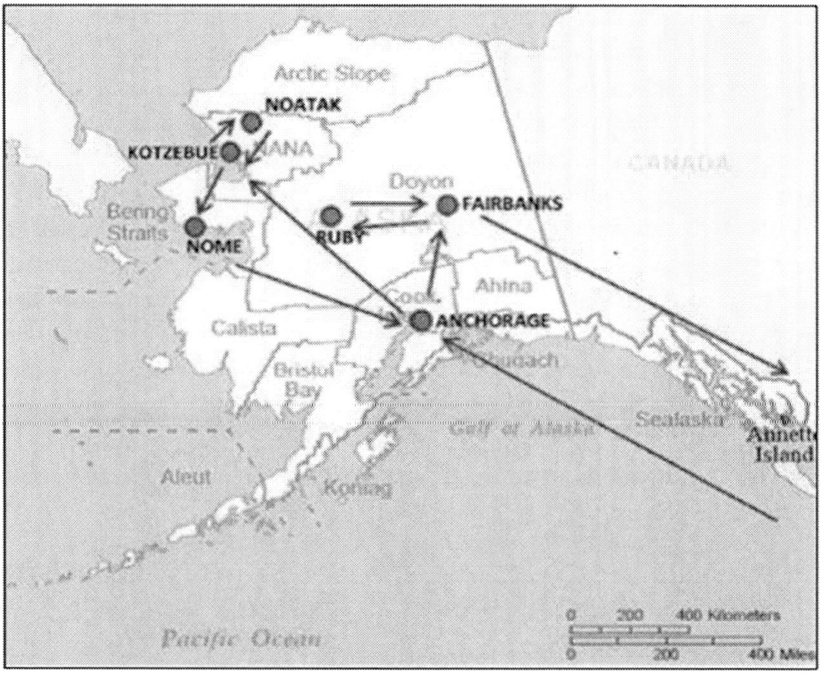

You can get there from here, just not easily.

In order to maximize the effectiveness of ONAP's reach and engage face-to-face with Native communities in Alaska, the most time and cost-efficient routing from the "lower 48" can look like a military transportation campaign. ONAP staff embarked on a mission that took them for separate consultation events, training seminars, meetings, and site visits via this aggressive itinerary:

- *Washington, DC to Anchorage*
- *Anchorage to Kotzebue*
- *Kotzebue to Noatak*
- *Noatak to Kotzebue*
- *Kotzebue to Nome*
- *Nome to Anchorage*
- *Anchorage to Fairbanks*
- *Fairbanks to Ruby*
- *Ruby to Fairbanks*
- *Fairbanks to Washington, DC*

In the Native Village of Kotzebue, the launching point for visiting north of the Arctic Circle, ONAP visited the husband-wife team that runs the local community radio station – KOTZ 88.9 FM, 720 AM – and brought back to the Commission a first-hand view of both the challenges and immense benefits associated with Tribal broadcasting. In the Native Village of Ruby, ONAP met with representatives of Yukon Telephone Company, the local telephone company, visiting their local home office and switching facilities and bringing back to Commission headquarters the perspective of a remote provider facing the challenges of the universal service reform era.

All of the outcomes of these valuable experiences – in Alaska and throughout Native America – and the relationships that were initiated and policy perspectives that were furthered, could only be accomplished when undertaken in face-to-face, on the ground, "muddy boots" efforts. These efforts profoundly affected ONAP's and the Commission's ability to provide fundamentally valuable input and, thereby, positively and proactively affected the quality of the Commission's important Tribal policy regulations and initiatives. As situations are evaluated and solutions identified, much more still remains to be done.

Many remote locations in Alaska can only be reached by air, boat, or barge. Such is the case for the tiny village of Noatak, located about 100 miles north of the Arctic Circle.

ONAP participated in a town hall meeting in Nome open to the general public. Record setting rainfall did not deter many residents from coming to the meeting. The event was covered by the local newspaper, the Nome Nugget, and by a local radio station.

CASE STUDIES

Introduction

The following case studies tie together all of the preceding sections of this Annual Report by illustrating the real world impact of the Commission's policies in Indian Country and in Native communities. While *output* reporting is important, ***outcome*** reporting that illustrates effects is better by an order of magnitude. That is, outcome reporting brings to life the relevance, importance, and real world impact of otherwise esoteric or complex Commission rulemakings and the true cost-benefit value of budget dependent Tribal consultation and outreach efforts. ONAP has selected certain Tribal Nation policy issues and highlighted their impact in Indian Country to make the words on the pages of this Annual Report come to life – all as part of ONAP's ongoing mission to bring Indian Country to the Commission and the Commission to Indian Country.

Universal Service

The Hopi Tribe

The Hopi Tribe is a federally-recognized sovereign Tribal Nation located in northeast Arizona. The 2,500 square mile Hopi Reservation is made up of 12 villages on three mesas, has a population of roughly 7,000 people, and is located within the Reservation of the Navajo Nation. Sparsely populated and geographically isolated, the Reservation faces daunting levels of unemployment and the nearest sizeable town is hours away. [56] In 2005, the Bureau of Indian Affairs reported that the Hopi Reservation suffered from 66 percent unemployment. Among those employed, 22 percent fell below the Federal Poverty Guidelines.[57] The Tribe's principal economic activities are tourism and agriculture. All government powers are vested in the Hopi Tribal Council, which is composed of a Chairman and a Vice-President, each serving four years, and Council members, who serve two-year terms. The current Tribal Council consists of 14 representatives from the villages of Upper Moenkopi, Bcavi, Kykotsmovi, and Sipaulovi. Representatives to the Council are elected either by a community election or by an appointment from the village kikmongwi, or leader.[58]

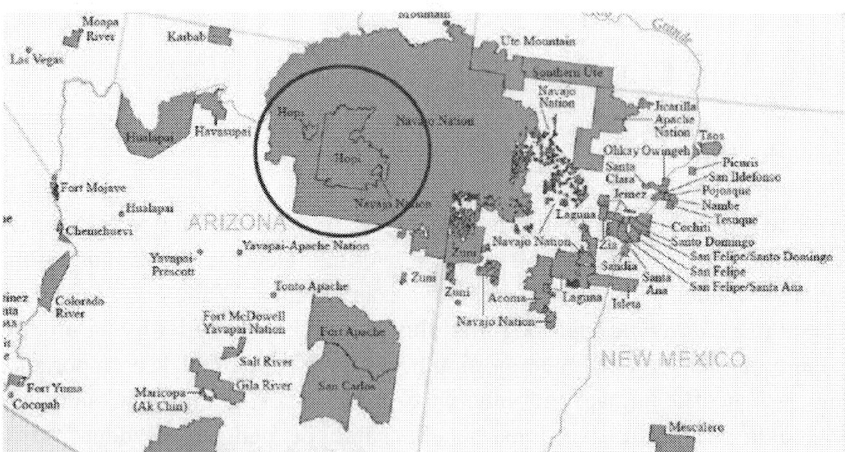

The economic and geographic challenges faced by the Hopi Tribe offered little financial incentive to provide telecommunications services to much of the Reservation. As a result, many people living on the Reservation lacked access to even basic telephone service as they entered the 21st century and, even where service was available, outages were frequent due to inadequate network capacity. [59] The Hopi Tribe determined that "[t]he lack of quality service was hindering the social and economic health of the Tribe."[60] So, in 2004, after years of inadequate telecommunications service on the Hopi Reservation, the Hopi Tribal Council established Hopi Telecommunications, Inc. (HTI) and required, among other duties, that HTI provide a telecommunications infrastructure that promotes economic development and Tribal sovereignty through empowerment, self-sufficiency, and self-regulation.[61]

HTI purchased the existing infrastructure from the previous wireline provider and began service in 2006.[62] Given the high costs of providing service in such a rural and remote area, coupled with the need to upgrade and expand the old copper infrastructure that it had purchased, HTI sought and received its ETC designation from the Commission and became the eighth Tribally owned ETC in 2007. HTI began receiving high-cost support in the first quarter of 2007. Today, HTI continues to play a vital role in a historically underserved area.

ONAP visited KUYI – 88.1 FM on one of its visits to the Hopi Reservation and participated in a live radio broadcast. Multiple volunteers at KUYI, which means "water" in the Hopi language, come from the nearby High School and take part in the award winning programming.

The Hopi Tribe was able to envision ownership and operation of its own telecommunications company in part because of the success of its long-standing radio station, KUYI – 88.1 FM. Licensed to the Hopi Foundation, KUYI is an integral part of the community on the Reservation and provided a community understanding and adoption platform for what HTI has been able to accomplish. Adoption can be driven by very different types of community anchor institutions in Indian Country, including the presence of previously well-adopted technologies, such as radio.

The Impact of Universal Service on Tribal Lands

The accomplishments of the Hopi Tribe and HTI, in the company's seven short years of existence, goes to the heart of universal service. Since its designation as an ETC, HTI has expanded and upgraded service significantly on the Hopi Reservation. For example, in 2008, HTI began offering its first DSL service.[63] Today, in addition to its wireline service offerings, HTI offers satellite phone and internet service for those who live beyond the reach of HTI's extensive upgraded copper network. HTI's broadband penetration rate today is approximately 40 percent. High-cost universal service funds have been used, and continue to be used, for the deployment, maintenance, and upgrading of critical infrastructure across the Reservation and are a cornerstone of HTI's goal to make broadband access available to all residents in its service area.

ONAP has visited the Hopi Reservation on two occasions – once in 2010 on the newly-formed team's first trip to Indian Country and again in 2012. ONAP had the opportunity to visit the two Tribally owned and operated communications companies on the Reservation – Hopi Telecommunications, Inc. and KUYI – 88.1 FM – Hopi Radio.

At HTI's main office, ONAP staff met with Carroll Onsae, HTI's General Manager and a member of the FCC-Native Nations Broadband Task Force, and discussed the impact of USF reform on HTI. At KUYI, ONAP staff was interviewed on a live radio broadcast during its visit to the station.

Both of these visits gave ONAP staff the opportunity to see firsthand the communications challenges that are experienced on the Hopi Reservation each and every day. ONAP was then able to bring that experience back to the Commission to inform policy decisions aimed at eliminating the digital divide in Indian Country.

The majority of HTI's residential subscribers – 78 percent – are eligible for Lifeline, [64] and today the company serves approximately 850 Lifeline households. [65] The Lifeline and Link Up programs have been vital assets as HTI expanded the reach and adoption of communications services across the

Hopi Reservation, providing cash flow for the growing company, making HTI's services affordable to its customers, and vastly increasing the telephone penetration rate. Since its creation, HTI has marketed and advertised its Lifeline offerings across its service area, even enlisting Tribal agencies in its outreach efforts. The Commission's decision to maintain Tribal Lands Lifeline support at its current support level and to preserve Link Up on Tribal lands was based in large measure on the experiences and comments of HTI, the nine other Tribally owned ETCs, and other carriers serving Tribal lands nationwide.

ONAP's visit to multiple Tribal Nations in Arizona included a visit to the offices of Hopi Telecommunications, Inc. and a visit to remote areas of the Hopi Reservation.

Another testament to HTI's skillful and determined efforts to expand the reach of communications services across the Hopi Reservation is the company's recent selection as a participant in the Commission's Lifeline Broadband Adoption Pilot Program. As one of 14 successful applicants – and one of two Tribally owned ETCs chosen to participate – HTI's application consisted of a comprehensive plan examining the effects of different subsidy levels and different equipment costs on broadband adoption.[66] The support available through the pilot program will help HTI determine how best to deploy broadband services to its Lifeline customers across the Hopi Reservation. In addition, HTI's participation in the pilot program will generate important data for the Commission regarding broadband adoption among low-income populations on geographically-isolated Reservations in high-cost areas. This represents yet another example of the partnership developed and born out of the Commission's ongoing government-to-government consultation with Tribal Nations – as articulated in the *Tribal Policy Statement*[67] – and, in this instance, with the Hopi Tribe concerning regulation of its Tribally owned ETC.

The Need for Tribal-Specific Rules

HTI and the Hopi Tribe have been invaluable partners to the Commission during the reform of the high-cost universal service program. Both individually and as a member of the National Tribal Telecommunications Association, HTI submitted comments, participated in meetings, and provided other invaluable first-hand information about the challenges associated with providing communications services on Tribal lands. Perhaps most importantly, HTI traveled to Washington to meet with Commission staff and, based on its own experience, explained why it was more expensive to provide communications services on Tribal lands. Under protective order, HTI literally opened its books to Commission staff and shared its financial information to help the Commission understand the increased costs of being an ETC on Tribal lands. Both through this effort and in its formal comments in the universal service reform docket, HTI illustrated the fact that requirements such as rights-of-way and permitting approval processes that are outside of a Tribe's control can cause considerable delays in deploying and upgrading infrastructure, which then result in higher costs.[68] HTI's contributions informed the Wireline Competition Bureau's adoption of a Tribal variable in its regression analysis for the calculation of High-Cost Loop Support, accounting for the increased costs of providing communications services on Tribal lands.[69]

The experience of the Hopi Tribe prior to the creation of HTI, as well as HTI's comments in the universal service reform proceeding, also informed the Commission's adoption of a Tribal government engagement obligation in its new Connect America Fund rules. The Hopi Tribe's decision to create its own telecommunications company to address the lack of service on the Hopi Reservation is illustrative of a situation common in Indian Country. That is, in many instances, industry alone cannot deploy service. Tribal Nations must coordinate with providers or, in some instances, make the decision to create their own providers. Carriers must coordinate with Tribal governments, as Tribal governments often are closest to the issues that present barriers to entry and can be the linchpins to potential economic solutions. Fostering more robust and substantive communication between Tribal governments and the ETCs serving on their lands is a fundamental goal of the new rules requiring engagement and the subsequent guidance released by ONAP, WTB, and WCB.[70]

The Hopi Tribe is also represented on the FCC-Native Nations Broadband Task Force by Carroll Onsae, HTI's General Manager since the company was established. Mr. Onsae's contributions, representing the perspectives of both a Tribally owned ETC and a rate-of-return carrier, have proven invaluable in matters before the Task Force, including universal service reform, the need for Tribal engagement rules, Lifeline reform, and access to spectrum over Tribal lands.

The role of the Hopi Tribe and its Tribally owned ETC – HTI – is but one example of the tremendous value that Tribal governments and providers bring to rulemakings across the Commission. Working sideby-side with ONAP, Tribal governments and providers exemplify what the Commission has so clearly articulated as a policy priority – "ensuring that Native concerns and voices are considered in all relevant Commission proceedings and initiatives."[71] While much has been accomplished since ONAP's creation, much more remains to be done in the months and years ahead.

Self-Provisioning and the Need for Spectrum

The Tribal Digital Village of the Southern California Tribal Chairmen's Association, headquartered on the Pala Reservation

The Pala Band of Mission Indians has over 900 enrolled members living on the Pala Reservation in Northern San Diego County. The Pala Reservation encompasses over 12,000 acres of land which, despite its relative proximity to

San Diego, a major metropolitan area, can be rocky, undeveloped, and mountainous. Because of these topographical challenges, until relatively recently, the Pala Reservation did not have access to terrestrial broadband services. This situation was not unique to Pala, but also similar to the lack of broadband access experienced in 16 other Tribal Nations that have Reservations in San Diego County. The challenge facing these Reservations in getting local telecommunications carriers to provide broadband was primarily economics. The cost of developing fiber in an area with topographical and terrain challenges could not be justified for the relatively small populations of the Reservations.

In 2001, the Southern California Tribal Chairmen's Association (SCTCA), an inter-Tribal government association consisting of 19 federally recognized Tribes located in Southern California, began developing, through its Tribal Digital Village project, a wireless broadband network with the help of a private "Digital Village" grant from Hewlett Packard under a collaborative arrangement with the University of California at San Diego. Utilizing a microwave backbone to connect 17 member Tribes in San Diego and Riverside Counties, the SCTCA built 21 backbone and relay sites, each supporting solar-powered antennas transmitting signal hops up to 26 miles apart, across a network that now extends over 450 miles.[72]

At initial deployment, the Tribal Digital Village system was designed to connect anchor offices and institutions to the Internet. These Tribal community institutions included Tribal government offices, public safety, health care, and educational facilities. This made the delivery of Tribal Temporary Assistance to Needy Families (Tribal TANF) through a shared services model much more available to the Tribal families in need of this important federal program, which is also an eligibility criterion for the Tribal Lands Lifeline and Link Up programs. As a credit to their ingenuity and illustrative of their determination, the Tribal Digital Village actually made broadband services available in some parts of the Reservations they cover before telephone service, water, or electricity was available.

Upon final deployment of the Tribal Digital Village, SCTCA and Pala Tribal government officials began to look at leveraging the connection to their anchor institutions to provide broadband services to residences on the Reservations, as well as to other Tribal buildings and facilitics. Dcmand for the service that the system brings is very heavy, and the Tribal Digital Village has completely maximized the viability and use of the unlicensed spectrum available to it. They began to make inquiries to the FCC and to the wireless industry community to investigate the possibility of obtaining an exclusive license or secondary markets agreement that would bring a solution to their spectrum needs. However, a new license was not available, and licensed spectrum suitable for robust wireless broadband use in and around the Reservation was not economically accessible via lease or other contractual arrangement with existing license holders. Because of the challenging geography and topology, trenching to lay additional fiber to homes and institutions was determined to be cost prohibitive, as their 2009 application for funding under the Recovery Act[73] was not approved in the second round of funding. The Tribal Digital Village has since had to retool and look to other viable options.

Led by Matthew R. Rantanen, IT Director of the Pala Reservation and the Director of Technology for the SCTCA, a wireless solution based on new "white spaces" spectrum technology for delivering broadband to residential and other resident-related facilities on the Reservation was planned and developed following the release of the Commission's *White Spaces Order* in September 2010.[74] Mr. Rantanen serves as a member of the FCC-Native Nations Broadband Task Force.

The Tribal Digital Village's Success at the Pala Reservation with White Spaces

Using the Tribal Digital Village network as their backbone, the Pala Reservation delivers wireless broadband service to homes and other Tribal buildings on the Tribal Digital Village network using commercial, off-the-shelf hardware. The Tribe erected a number of intermediate distribution towers in order to deliver broadband, via white spaces, to the home. Already, 10 percent of Pala residences are paying subscribers on the TDVNet, the consumer arm of the Tribal Digital Village.

The promise of the white spaces solution in the Pala Tribe's community may also provide potential economic opportunity, as the Pala Tribe and Tribal Digital Village are exploring ways to augment their subscriber base to include non-Native residents who reside in nearby off-Reservation areas. In these areas, as with the Pala Reservation before the launch of the Tribal Digital Village, broadband is not readily available today.

The Tribal Digital Village Builds a Local, Native Ecosystem

The Tribal Digital Village was originally developed between 2001 and 2003 to link unserved Reservation communities over a network spanning 350 miles via fiber and wireless connections to Tribal administration buildings, fire stations and other public safety facilities, utility departments, schools and libraries, EPA departments, and Head Start programs.

Once deployed, the network grew to be an incubator of technological expertise for Native youth. The Shadow Project provides training and mentoring for young people interested in technology and networking. Courses in video and audio production are offered in a media studio located on the Pala Reservation.

The Tribal Digital Village now links 14 Reservations in San Diego County and three Reservations in Riverside County through a network that has grown to more than 450 miles. Homes and businesses are now served as well.

Courses to assist in developing digital literacy among Tribal members are numerous and tied to the operation and success metrics of the Tribal Digital Village. Recognizing the digital divide in nearby off-Reservation areas, these courses are also available to non-Native members of the local community.

Coordinating closely with SCTCA leaders and the Tribal Digital Village management, ONAP has monitored the ongoing development of the Tribal Digital Village and the Pala Tribe's use of white spaces to deliver broadband to previously unserved homes and community facilities in and around the Pala Reservation. ONAP's experiences consulting with the Pala Tribe and the SCTCA – along with site visits to the Tribal Digital Village's networking, tower, and system sites – have helped inform the Commission on an important model and success story of a Tribal government self-provisioning initiative that is bearing fruit for multiple Tribal Nations.

During ONAP's 2012 Tribal Broadband and Telecom 101 training and consultation workshop held on the Pala Reservation, ONAP managers, in coordination with Tribal Digital Village and Pala Tribe officials, adjusted the agenda in progress in response to the interests of the attendees from other Tribal Nations in touring the Tribal Digital Village's facilities and receiving a hands-on briefing from Mr. Rantanen. Monitoring how individual Tribal Nations are successfully finding solutions to the lack of broadband and telecommunications services, cross-pollinating the information to other Tribal Nations who may benefit from lessons learned, and bringing this invaluable information back to the Commission is one of ONAP's core missions.

The use of white spaces to deliver broadband services to members is a prime example of spectrum self-provisioning efforts gaining in interest in Indian Country. Other impressive examples of Tribal Nations self-provisioning wireless solutions in unregulated spectrum-based or hybrid models can be found at the Coeur d'Alene Tribe in northern Idaho and the Yurok Tribe in northern California. Spectrum needs are particularly great because the Reservations of many Tribal Nations are located in rural areas with challenging terrain, exemplified by the badlands of the northern plains and the mountainous forests of the Pacific Northwest. In areas similar to these, laying fiber across the existing geography is not presently feasible. Coupled with the general lack of available spectrum and, often, the lack of interest of existing licensees to develop broadband on Tribal lands, some Tribes have successfully pursued their own solutions, as in this case with the Pala Band of Mission Indians and the Southern California Tribal Chairmen's Association.

The Important Need for Tribal Nation Access to Licensed Spectrum – White Spaces is Just One Potentially Temporary Solution

The need for more access to robust licensed spectrum to serve Indian Country is not just a critical need but, potentially, a resource that can mean the difference between life or death. A white spaces or unregulated spectrum solution will not be technically feasible in all situations in Indian Country, nor will it necessarily bring the assurance of longevity in growth for new uses, such as in the case of the Tribal Digital Village. Spectrum needs, topography, resource availability, and existing states of deployment vary widely among the 566 federally recognized Tribal Nations, as well as on the Hawaiian Home Lands. In many parts of Indian Country where a white spaces solution simply will not work, access to traditional, licensed spectrum is an issue that touches many facets of daily life, including public safety and economic development. In any number of Reservations, huge coverage holes exist – including residential areas and along isolated and desolate roads, where a simple car malfunction or accident can have serious consequences because of an inability to call for help. As a result, at one level or another, everyday lives are at risk in Indian Country.

Since the Commission began its Tribal outreach in 1999, Tribal Nations have illustrated their needs for spectrum-based services. In very positive and encouraging recent developments, interactions with Tribal leaders and major inter-Tribal government and economic development organizations illustrate an ever burgeoning awareness and interest in developing greater utilization of spectrum over Tribal lands.[75] As a direct result of ONAP's efforts in coordination with individual Tribal governments, ONAP's hosting of the Tribal Broadband and Telecom 101 training and consultation workshops, and ONAP's work with the FCC-Native Nations Broadband Task Force, Indian Country is more engaged on this critical issue than ever before. NCAI is building a campaign among and for Tribal Nations to build a more detailed and factual record in the *Spectrum Over Tribal Lands NPRM* proceeding. Virtually every corner of Indian Country is involved, as comments and letters received in the proceeding include those from NCAI, SCTCA and the Tribal Digital Village, the National Tribal Telecommunications Association, Native Public Media, the Confederated Tribes of the Colville Reservation, and the Navajo Nation Telecommunications Regulatory Commission.

Another indicator of manifest interest in direct licensing and new regulatory actions for Tribal Nations was routinely found in the Native Learning Lab at each of ONAP's broadband and telecommunications training and consultation workshops. Training and demonstration sessions on the Spectrum Dashboard have been among the most lively and well-attended of the training modules at the Native Learning Lab. Attendees were usually surprised at the number of licenses over Tribal lands in spectrum bands suitable for wireless broadband services. For a number of Tribal workshop attendees, the Dashboard's ability to provide licensee and band information represents the first step for individual Tribes to reach out to licensees and seek leasing, partnership, or other arrangements which will ultimately result in the provision of service over Tribal lands.

Finally, the upcoming Tribal Mobility Fund Phase I auction, expected in 2013, has generated great interest in spectrum availability from Tribes who see the auction as an opportunity to help fund build out of 3G and 4G service in locations where no such service exists. Because participants in the upcoming auction, limited to ETCs serving Tribal lands, must come to the auction with access to spectrum, ONAP anticipates a marked interest in the general area of spectrum management. This is why, as noted earlier in this report, the Tribal Mobility Fund is one of ONAP's top consultation priorities for 2013.

Ultimately, the Pala Band of Mission Indians, SCTCA, and the Tribal Digital Village is an excellent example of a Tribe, an inter-Tribal government association, and a Tribal business that were able to leverage white spaces to self-provision a wireless broadband service, for the time being. However, much of Indian Country still has yet to realize the promise of broadband. ONAP will continue to work and consult with Tribes making their way to deployment as well as collaboratively help inform Commission policy making efforts on matters which directly impact wireless broadband and telecommunications deployment on Tribal lands.

Tribal Engagement

The Confederated Tribes of the Colville Reservation

The Confederated Tribes of the Colville Reservation homeland is roughly 2,100 square miles of Pacific Northwest land in eastern Washington State. The eastern and southern border of the Reservation is the mighty Columbia River, one of the great historical trade route rivers of Northwest Indian Country. The Colville Reservation also has steep mountain ranges, valleys limited in vehicle access, and old-growth, undeveloped timber land. With a population of over 9,300,[76] residents live in communities spread throughout the Reservation. It is the home to the descendants of 12 separate indigenous Tribal peoples, including the last homeland of Chief Joseph, the famous Nez Perce leader.

The Colville Confederated Tribes have focused their governmental attention on the improvement of their communications for well over a decade, actively participating in both federal and state regulatory proceedings, addressing issues to local and national carriers, applying for federal programs that address critical infrastructure needs, and dedicating Tribal leaders and resources to working with the Commission on multiple formal advisory committees, including the Consumer Advisory Committee (CAC), the Intergovernmental Advisory Committee (IAC), and the FCC-Native Nations Broadband Task Force. Former Colville Business Council Vice-Chairman John Stensgar (CAC), former Council Member Cherie Moomaw (IAC), and former Council Member Susie Allen (FCC-Native Nations Broadband Task Force) have all spent time working with the Commission in hands-on policy development work.

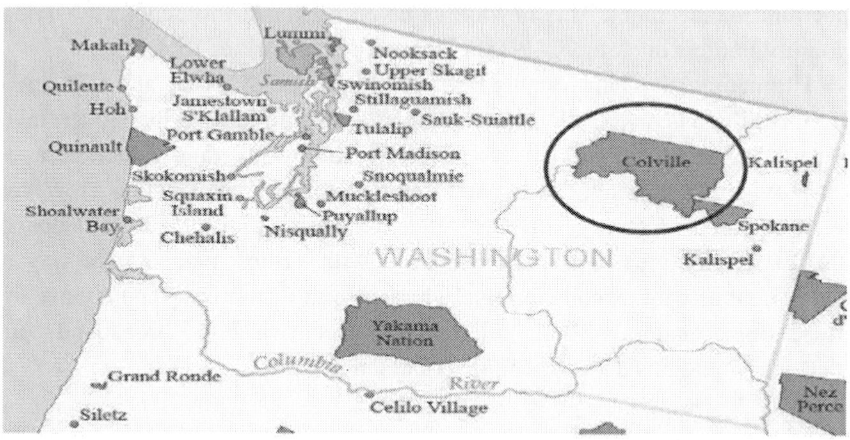

The Colville Confederated Tribes also has an IT Department that has actively worked to address the lack of telephone and Internet access throughout the Reservation. They have confronted many regulatory, economic, and geographic obstacles, and now operate some of the Tribal Nation's own infrastructure to serve previously completely disconnected communities. The topographical features of the Reservation, coupled with the fact that residential communities are widespread within the Reservation's borders, presents great challenges. Large areas of the Colville Reservation completely lack mobile phone coverage, and several of those areas are also residential areas populated by Tribal members. Laying fiber is extremely difficult and expensive in some of the remote portions of the Reservation, as is the placement of mobile wireless communications towers in mountainous, heavily forested terrain miles from the nearest power source.

Residents of the Reservation routinely drive to places, exact locations actually, within the Reservation where wireless signals are available. While the signals in these locations are not robust, at least residents can actually make a wireless telephone call. The impact of this situation during an emergency, when mere minutes can mean the difference between life and death, is frightening to imagine. It is a reality for the residents of the Colville Reservation. In the event of a breakdown or an automobile accident, they also assume great personal risk traveling in areas of the Reservation where help is not available due to a lack of wireless services. Overall, when help is not "just a call away," the impact on public safety is profound.

Perhaps the most detrimental impact of the spotty nature of connectivity to wireless phone networks and terrestrial broadband networks on the Colville Reservation is the significantly negative effect on the Tribal Nation's and Tribal members' ability to generate economic progress, leverage off-Reservation resources in education and health care, and effectively manage growing Tribal government operations.

ONAP Consultation and On-The-Ground Findings Confirm Issues

The challenges faced by the Confederated Coville Tribes have been a focus of the Commission for nearly a decade, and Commission staff have made multiple visits to the Colville Reservation, undertaking specific outreach and training efforts there. The Commission's first visit to Colville was in 2003 to meet the Tribal government. In 2004, the Commission's Office of Intergovernmental Affairs (pre-ONAP) hosted an ITI Regional "Train the Trainer" session at the Nespelem Tribal Longhouse, the traditional meeting place for the community of the Nespelem District of the Tribal government. In 2004, the Nespelem Tribal Longhouse had, literally, the last telephone for many dozens of miles into the interior of the Reservation. At that time, the entire Reservation community of Disautel, approximately 30 miles northwest of Nespelem, was completely without telephone service. The Tribe eventually addressed that situation themselves.

Mobile phone service within the boundaries of the Colville Reservation is spotty, with significant dead zones in and around residential areas.

During a 2012 visit, ONAP staff met Matt Haney of the Colville Tribal Police Department. Officer Haney provided a realworld example of how the lack of mobile phone service impacts public safety.

A patrol officer had found some unexploded ordinance within the Reservation. He needed to call the situation into his dispatch desk for additional support to, among other things, cordon off the area to keep people safely away and control traffic flow.

However, because so many Tribal members have police scanners in their homes for social and recreational use, using the police radio to transmit information about the unexploded ordinance would have actually brought in more curious people who heard about it on their scanners rather than keep people away.

"We really needed to be able to use a cell phone in that instance to call headquarters," Officer Haney said.

With the creation of ONAP, on two separate visits, one in 2011 and one in 2012, the Commission resumed efforts to work on the ground in consultation with the Colville Tribal government. In 2011, ONAP met with the Colville Business Council and saw portions of the Reservation few federal officials have ever seen, including the top of Omak Mountain – where, at the end of rough mountain trails next to steep ravines, stood several tall towers laden with communications equipment and assets that serve primarily off-Reservation consumers. In 2012, the Colville Confederated Tribes hosted extensive and detailed meetings with Tribal leaders and members of Tribal government departments, including management and planning, IT and telecom, economic development, public safety, and fish and wildlife. ONAP has thus developed a far greater understanding of the challenges associated with communications deployment on the Colville Reservation.

During its 2012 visit, ONAP managers and staff were led by members of the Tribal government managerial staff on a tour of areas of the Reservation of particular need or priority to the Colville Tribal government. We walked along a route where a carrier had reported broadband service via fiber on the National Broadband Map, yet saw none. This situation impacts the Tribe's – or any other carrier's – ability to apply for telecommunications build-out and deployment grant programs, as unserved portions of the Reservation were being incorrectly shown as having access to broadband. Conversely, we were then shown a significant presence of dark fiber that was serving no Tribal residents and providing no broadband service. The fiber was actually placed in the bottom of a ditch and extended several miles, unconnected at either end to any service. Moreover, some of the dark fiber reported by the incumbent local exchange carrier as being deployed was not buried and lay in pieces on the side of the road.

ONAP traveled to the top of 5,747-foot Omak Mountain, located on the Colville Reservation in eastern Washington state, to see towers and communications infrastructure and equipment that serve primarily off-Reservation consumers.

Dark fiber lies unconnected on the side of the road within the borders of the Colville Reservation. For Tribal members, the long stretch of dark fiber running within a roadside ditch – displayed on maps as actively supporting deployed telecommunication services – was a great source of frustration.

ONAP met with the entire Colville Business Council at the Confederated Colville Tribes Headquarters to discuss communications infrastructure needs.

Tribal Council Members and Tribal staff managers noted a lack of productive basic communication between the incumbent carrier and the Tribe, resulting in basic misunderstandings over simple things which formal meeting and routine coordination could have fixed. Tribal officials noted certain objectionable actions by incumbent carriers, including entering onto Tribal property without notice and without permission to place carrier signage on sheds, fences, and buildings—pursuant to Tribal permitting requirements. Similarly, ONAP heard reports of the carrier trenching along a roadside without notification to the Tribe.

Tribal Engagement – Future Promise of Better Results for the Colville Reservation

ONAP's 2012 meeting with Colville Tribal government officials included a detailed review of how the Tribe felt about the quality of service provided by carriers serving the Reservation. This discussion then dovetailed into a detailed and productive training session on the new Tribal government engagement requirement provisions of the Connect America Fund regulations. Part of ONAP's responsibility is to work with Tribal governments and carriers to help both parties be better prepared to identify and address specific issues with data, problem descriptions, success metrics, and delineated potential and practicable outcomes.

To further help prepare the Colville Business Council for meeting with carriers, ONAP staff discussed strategies that the Tribal Nation might consider, in an effort to make engagement meetings with carriers productive and address ongoing deployment concerns. For example, ONAP staff recommended that the Tribe prepare for its Tribal engagement meetings based on a multi-faceted approach rooted in areas of concern articulated by the Tribal Nation. For the Colville Reservation, these issues include public safety, as well as Tribal notification requirements and processes associated with trenching, erecting, and placing signage on equipment on Tribal lands. These issues also include a joint Tribe and carrier understanding of Reservation areas that are lacking in service, discussions about accurate reporting of services provided, and future deployment planning to presently fill gaps – to eventually but deliberately bring advanced wireless services to all parts of the Reservation.

While only an element of pessimism lingered, explained as a result of multiple past failed attempts to engage carriers, several members of the Colville Business Council and management staff were optimistic that the new opportunities to work together in the upcoming first round of Tribal government and carrier engagement meetings would be productive, collaborative, and solution-oriented. They conveyed their hope and anticipation that these carrier and Tribal engagement sessions would provide the formal structure and framework for Tribal Nations and carriers to work together and avoid misunderstandings on day-to-day issues, providing a foundation for future shared vision and solutions.

Ongoing Training and Consultation on this Issue with Tribes

The Tribal government engagement obligation is one of ONAP's top Tribal consultation priorities for 2013. Continuing to coordinate with the multiple communications companies that serve Tribal lands also continues to be a top priority for ONAP. Because the Tribal engagement requirement is a relatively new regulatory requirement, ONAP will coordinate with Tribal governments and carriers nationwide who serve Tribal lands, in individual and shared settings, to get a sense of their experiences during this first year. ONAP will then compile specific recommendations to make future Tribal-carrier engagement meetings more productive, facilitate more meaningful engagement meetings between Tribal governments and ETCs, and begin to develop the best practices that ONAP must produce under Commission order.[77] All of this will be accomplished both through individual consultations with Tribal governments and meetings with carriers, and in the context of regional trainings and Reservation visits.

ONAP's consultation mission, on behalf of the Commission, is one that Tribal governmental officials across the country have genuinely and enthusiastically supported. Support for the Commission's consultation process thus far has taken the form of official resolutions from the National Congress of American Indians, Tribal leader feedback received while engaging in our mission on Tribal lands, and comments by Senators and members of Congress during multiple hearings at ONAP-presented testimony before the Senate Commerce and Indian Affairs Committees and the House Subcommittee on American Indian and Alaska Native Affairs. Most recently, on December 5th, at the White House Tribal Nations Summit, during a listening session that included seven separate agencies, the most senior Tribal leaders from Tribal Nations in New York, California, Washington, Arizona, and Oklahoma stood and expressed their appreciation for the Commission's commitment to coordinate with Indian Country through the successful ongoing approach of ONAP's work. The future of the Commission's Tribal consultation and coordination policy will build on its initial success, leveraging what worked well in 2012 and before by continuing the forward momentum and, among other efforts, increasing the amount of legal and technical support and training provided to Tribal governments. Much progress has been made, but much more progress remains ahead of us.

Radio Broadcasting and the Tribal Priority

The Promise of the Tribal Priority

The Commission has a longstanding commitment to promoting Tribal self-sufficiency and economic development, as well as providing access to communications services to Tribes.[78] This is especially true in the context of broadcast radio. In 2009, the Commission recognized that, despite the existence of a number of Tribally-owned radio stations, the traditional allocation priorities had not fulfilled the Section 307(b) mandate to "make such distribution of licenses ... among the several States and communities as to provide a fair, efficient, and equitable distribution of radio service"[79] with respect to Tribal lands.[80] In a series of orders predicated on the federal trust relationship and promoting Tribal sovereignty, the Commission amended its rules to enhance the ability of federally-recognized Native American Tribes and Alaska Native Villages "to receive radio service tailored to their specific needs and cultures" and "to increase ownership of such radio stations by Tribes and Tribally-owned entities."[81] These orders were the result of consultation with and comments from Tribes. The Commission created the Tribal Priority to enhance and promote the economic and cultural values of Tribes.[82] The most recent application of the Tribal Priority empowers Tribes to more readily obtain full power commercial radio licenses.

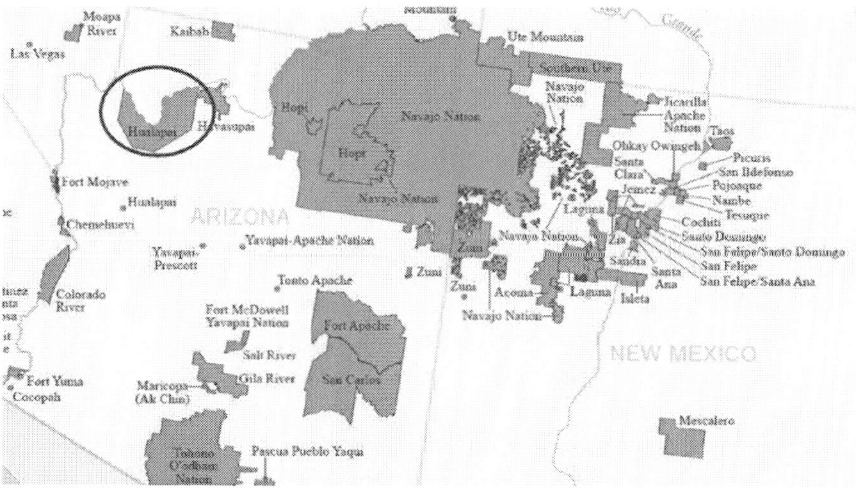

The Hualapai Tribe and Its Goal of Having a Radio Station

The Hualapai Indian Reservation, home to the Hualapai Tribe, is in northwest Arizona, encompassing nearly one million acres along 100 miles of the Colorado River and the Grand Canyon. The Reservation is shaped like the letter U and sits on the southern side of the Grand Canyon, stretching to the Grand Canyon's western corridor and Lake Mead. The topography varies greatly, from dry grasslands to cedar and pine forests to the Grand Canyon. Elevations range from 1,500 feet at the Colorado River to over 7,300 feet at the highest point on the Reservation.[83] The Tribal Nation's headquarters is located in the Reservation's largest town, Peach Springs. More than 1,600 people reside on the Hualapai Reservation,[84] with nearly 1,300 in Peach Springs. Outside of Peach Springs, therefore, the Reservation is sparsely populated.[85]

The Hualapai Tribe is governed by a Tribal Council, consisting of a Chairperson, a Vice Chairperson, and seven other Council members. Currently, the Hualapai Tribe faces severe economic circumstances.

Unemployment hovers around 50 percent and the principal economic activity is tourism.[86] To combat this endemic poverty, in 2007, the Hualapai Tribe opened the Grand Canyon Skywalk, a major tourist attraction in western Arizona. Currently, there is no broadcast radio service across much of the Reservation, including on isolated roads. This poses a significant public safety risk. On October 11, 2012, in an effort to increase the level of communications services on the Reservation, the Hualapai Tribe filed the first petition for rulemaking from a Tribal Nation itself using the Tribal Priority.[87] The Hualapai Tribe petitioned the Commission to amend the FM Table of Allotments to add a new channel, the first Tribally-owned commercial facility, based in Peach Springs.[88] In its petition, the Tribe showed that the proposed primary contour would cover a total area of 816 square kilometers, with nearly 85 percent of that coverage on Tribal land.[89] A radio station on the Hualapai Reservation would provide both an economic boost to an impoverished area and allow the Tribe to follow in the path of providing uniquely-tailored cultural programming to Tribal members. This valuable resource, which would be the first Tribally-owned service on the Reservation, would help preserve Native language, cultural values, and community for the Hualapai Tribe. On December 21, 2012, the Audio Division issued a *Notice of Proposed Rulemaking* to amend the FC Table of Allotments, seeking comment on the Hualapai Tribe's petition to use the Tribal Priority for the proposed allotment.[90] On March 1, 2013, the Audio Division granted Hualapai's petition for rulemaking and allotted FM Channel 265A at Peach Springs, Arizona, as a Tribal Allotment, one of the first two full power commercial FM allotments under the Tribal Priority.[91]

The Hualapai Indian Tribe is the first Tribal Nation to seek the Tribal Priority, and the second applicant after the Navajo Technical College, an entity of the Navajo Nation. In October 2012, ONAP began consultations to discuss the operation and potential of the Tribal Priority with the Hualapai Tribe's "radio task force," which consists of the Vice Chairman of the Tribe, a Tribal Council member, the manager of the Tribe's online radio station, and their engineer. The request for consultation was initiated by a Hualapai Tribal Council Representative at the 69th Annual Conference of the National Congress of American Indians in Sacramento. This type of government-togovernment consultation is one of the fundamental aspects of ONAP's mission meets the Commission's mandate in the Tribal Priority rules. It represents an important aspect of ONAP's consultation and coordination field work with Tribal Nations.

The Hualapai Tribe already has a media presence with its own online radio station, EPHC, and the Tribal Council initially considered a low power FM station several years ago, but determined that the coverage of such a station would be inadequate for the community.[92] While learning about the new rules, the Tribe considered the new Tribal Priority for full power commercial licenses as an opportunity to procure a valuable economic asset that could also be used to promote cultural values and language.

The Hualapai Tribe quickly recognized that, with these new rules, the Commission had acted to protect the Tribe's assets and the economic value of the Tribal Priority in a commercial FM context by adopting the threshold qualification window process.[93] The Tribal Priority effectively empowers Tribes to obtain full power commercial radio licenses without having to purchase the license in a costly competitive auction, a process which could not ensure their ability to broadcast to their own Tribal Nation's communities.

After petitioning the Commission for rulemaking, the Hualapai Tribe initiated a request for consultation on the procedures and operation of the Tribal Priority rules. ONAP and the Audio Division of the Media Bureau were directed by the Commission in its Report and Order creating the threshold qualifications window process to coordinate with Tribes to ensure that any questions involving the Tribal Priority's threshold qualification and licensing processes are answered.[94] In response to the Tribal governmental request, ONAP staff met with the Tribe's "radio task force" to discuss the communications needs and priorities of the Tribal Nation with regards to their petition. ONAP and the Audio Division subsequently coordinated closely on the petition, meeting the Commission's expectations for Tribal consultation. During these discussions, the Tribal leaders discussed the importance of Hualapai voices speaking the Hualapai language on air, the economic value to the Tribe of a full power commercial FM station, the public safety aspects in times of emergency on the Reservation, and their desire as a government for a license area to cover the planned community and economic venture growth of the Hualapai Tribe.

The Voice of Tribal Radio and the Potential for Growth

As the number of Tribal voices on radio has increased, the value of Tribal radio broadcasters to empower Tribal Nations and Tribal peoples and to promote Tribal sovereignty becomes more and more apparent. Today, there are Tribal Nation and Native community radio stations throughout the United States. Native Public Media, the association of Tribal Nation and Native community radio stations, lists 55 stations in their database, ten of which are streaming their services online.[95] Since its creation, ONAP has made it a priority to visit several of these stations, including KOTZ - 89.9 FM or 720 AM of Kotzebue, Alaska; KIDE - 91.3 FM of the Hoopa Valley Indian Tribe, in northern California; KOPA - 91.3 FM of the Pala Indian Tribe in southern California; KWSO - 91.9 FM of the Confederated Tribes of Warm Springs Reservation, in Oregon; KCUW-LP - 104.3 FM of the Confederated Tribes of the Umatilla reservation, in Oregon; KWIS - 88.3 FM of the Coeur d'Alene Tribe, in Idaho; KUYI - 88.1 FM of the Hopi Tribe, in Arizona; and WOJB - 88.9 FM of the Lac Courte Oreilles Band of Lake Superior Anishinaabe, in Wisconsin.

The role of Tribal radio is unique to Indian country. As the Commission formally recognized over a decade ago, Tribal governments have the right to set their own communications priorities and goals for their citizens and communities. Each station addresses the unique needs of the Tribal Nation or entity that owns it, from broadcasting emergency messages in areas with poor phone service, to preserving and promoting Native languages and culture, to promoting economic self-determination. Foundational to the purpose of many Tribal broadcasters is filling their rightful place in the media landscape, where the popular dominant culture perceptions of the indigenous nations and peoples of this country have often been formed by Hollywood movies, old west television shows, and college mascots that wholly inaccurately portray Native Americans. Telling one's own story, broadcasting in one's own voice, in an exercise of self-determination and self-reliance, is so important a goal of so many broadcasters in Tribal communities that its value cannot be overstated.

In spite of the increasing number of Tribal radio stations, however, their ranks are still very low relative to the number of Tribal Nations and communities of Native populations. That is, despite comprising 1.5 percent of the country's population, less than one-third of one percent of radio licenses are Tribal Nation and Native community owned stations.[96] Tribal Nations across the country and national inter-Tribal government and industry organizations, such as the National Congress of American Indians and Native Public Media, have articulated the need for many more Tribal voices in broadcast radio. The Commission's partners in Indian Country helped it create these rules, and stand ready to help ONAP and the Commission assist and inform Tribal Nations.

ONAP visits to Native radio stations often include on-air interviews, which occurred during this visit to KNBA 90.3 FM in Anchorage, AK. KNBA is operated by the Koahnic Broadcast Corporation, a non-profit, Alaska Native governed and operated media center.

The Hualapai Tribe's application for the Tribal Priority is their groundbreaking first step into the world of radio broadcasting. Every week, Tribal government and development representatives contact ONAP and the Audio Division to learn more about their opportunities under the Commission's Tribal priority radio broadcast rules. ONAP and the Audio Division, on behalf of the Commission, have much more good work to do in making Tribal Nations aware of, and assisting them with, this new opportunity.

In these efforts, ONAP and the Audio Division enjoy an excellent and mutually beneficial working relationship, illustrative of the potential across all the Bureaus and Offices. Audio Division managers and staff participated in five training and consultation events hosted by ONAP in 2012, contributing expert presentations on panels and meaningful guidance in side or "add-on" meetings with individual or groups of Tribal Nations. Both offices are dedicated to seeing the successful operation of these rules, and their responsibilities to Tribal Nations as federal offices, come to fruition in the deployment of services on Tribal lands. The Audio Division even prepared diagrams and information pieces, in coordination with ONAP, that assist immeasurably in building awareness and understanding of the Tribal Priority in radio broadcast.

Eligible Telecommunications Carriers

The Standing Rock Sioux Tribe

The present day homeland of the Standing Rock Sioux Tribe is the Standing Rock Reservation, which encompasses approximately 2.3 million acres on the borders of North Dakota and South Dakota. The eastern border of the Reservation is the Missouri River, the historically sacred river of so many indigenous peoples of the Great Plains. The Standing Rock Sioux Tribe's experience in the communications field is a proud one, exhibiting their dogged determination – Tribal self-determination – and their self-reliance in the face of extreme challenges. Standing Rock experiences an unemployment rate of 79 percent,[97] an average annual income of $10,000, and a high school graduation rate of 60 percent or less.[98] With a population of just over 6,000 residents, the Reservation averages 0.4 persons per square mile. Agriculture is the primary Reservation industry; however, this sparsely populated Reservation experiences temperatures in the winter from 30 below zero to 17 above zero with routine severe blizzard conditions, and temperatures in the summer from 60 to 110 degrees, with occasional drought.[99]

The Standing Rock Sioux Tribe operates under its Tribal constitution of April 24, 1959. The Tribal Nation is governed by a Tribal Chairman, Vice-Chairman, Secretary, and 14 Council Members. Certain Tribal Council Members occupy at-large positions and are elected by the entire Tribe, while other Council Members represent the eight Districts of the Tribe and are elected from their Districts.[100]

For nearly 20 years, the Standing Rock Sioux Tribe has endeavored to solve its communications problems. This involved multiple unsuccessful attempts to attract deployment and coordinate with local incumbents, both wireline and wireless, and even involved the failed sale of exchanges on the Standing Rock Reservation to an ETC owned by a neighboring Tribal Nation – the Cheyenne River Sioux Tribe Telephone Authority, the oldest Tribally owned wireline ETC. A refocused Tribal Council began to look at the possibility of wireless services from carriers who held licenses on the Reservation, but had not yet built out. In these many endeavors, the challenges of cost, jurisdiction, geo-political factors, and lack of attention conspired against the Tribal Nation's planning. Over the course of these experiences, the focus of the Standing Rock Sioux Tribe narrowed again and become even more resolute.

Today, there are ten Triballyowned ETCs:

- Cheyenne River Sioux Tribe Telephone Authority
- Fort Mojave Telecommunications, Inc.
- Gila River Telecommunications, Inc.
- Hopi Telecommunications, Inc.
- Mescalero Apache Telecommunications, Inc.
- Saddleback Communications (owned and operated by the Salt River Pima Maricopa Indian Tribe)
- San Carlos Apache Telecommunications and Utilities, Inc.
- Standing Rock Telecommunications, Inc. (the only Tribally-owned wireless ETC)
- Tohono O'odham Utility Authority
- Warm Springs Telecommunications Company

In an effort to address the appalling state of communications services on the Reservation, the Standing Rock Sioux Tribe created Standing Rock Telecommunications, Inc., (Standing Rock Telecom), a Tribal Nation-owned and operated commercial mobile radio service (CMRS) provider, to provide mobile voice and data services within the entire Reservation. Standing Rock Telecom gained access to spectrum through a secondary market agreement negotiated with a major wireless carrier, and sought ETC designation from the Commission.

In a two-part decision, the Commission designated Standing Rock Telecom as an ETC throughout the Reservation, basing its decision both on the historical trust relationship the Commission shares with federally recognized Tribes and the Commission's commitment to promote the availability of affordable communications services to underserved consumers, many of whom reside today on Tribal lands.[101] This decision was groundbreaking in at least two respects. First, it marked the first time that the Commission designated a Tribally owned ETC to serve entirely within the exterior boundaries of its Reservation, recognizing the population of the Reservation as the Tribe's community, including partial rural wire centers.[102] Second, the Commission concluded that redefinition of the underlying rural telephone company study areas was unnecessary, and therefore no state commission consent was necessary, because the Commission designated Standing Rock Telecom throughout the entire rural service area within the Standing Rock Sioux Tribe's Reservation – the full extent of the Commission's jurisdiction.[103]

Standing Rock Telecom, as the first Tribally owned wireless ETC, thus became eligible for much-needed universal service support, including high-cost and low-income support, to continue the process of extending service throughout the Reservation.[104] In September 2011, Standing Rock Telecom participated in the Commission's first reverse auction, held as part of Phase I of the Mobility Fund. In October 2012, the Wireless Telecommunications Bureau announced that Standing Rock Telecom was one of the auction's successful bidders, resulting in $3.3 million in total assigned support and coverage for up to 1,290 road miles on the Standing Rock Reservation.

At the invitation of the Tribal Council, the Commission first visited the Standing Rock Reservation in 2004, represented by managers and senior staff of CGB and the Office of Intergovernmental Affairs. These Commission officials even held a call-in show on the Tribal radio station located on Standing Rock, KLND - 89.5 FM, which serves both the Standing Rock and Cheyenne River Reservations and surrounding areas. Very quickly the

Commission learned firsthand how dedicated the Tribe and its people were to improving their telephone and broadband situation. It is noteworthy that the radio station's name in the Lakota language is Wolakota Wiconi Waste, meaning "through unity a good life." During the course of their work with the Commission to address their communications challenges and creating Standing Rock Telecom, improving the quality of life on the Standing Rock Reservation has been a foundational goal and a constant message of the Tribal Nation. Chairmen, Council Members, and senior management officials of the Standing Rock Sioux Tribe have met with FCC Chairmen, Commissioners, Bureau Chiefs, and staff in multiple Bureaus. The work between the Commission and Standing Rock has epitomized the many valuable opportunities found in the unique nation-to-nation relationship that the Commission and Tribal Nations share.

ONAP visited the Standing Rock Reservation in 2011, not long after Standing Rock Telecom' s groundbreaking ETC designation. ONAP staff met with Chairman Murphy of the Standing Rock Sioux Tribe, the management of Standing Rock Telecom, and other Tribal government officials.

Having the opportunity to experience first-hand the extremely remote nature of the Reservation and the telecommunications challenges faced by both the Tribe and its newly-designated ETC proved invaluable back in Washington, DC, as ONAP coordinated across the Commission on universal service reform and its impact on Tribal lands. Nothing substitutes for face-to face contact and consultation in federal-Tribal government-togovernment affairs – particularly in situations in which the conditions are dire.

The Importance of the Opportunity for Tribal Nation Self- Provisioning

Standing Rock illustrates quite dramatically a number of important Tribal Nation policy initiatives at the Commission. First and foremost, Standing Rock personifies the importance of affording opportunities for Tribal Nations to self-provision communications services. Facing a virtual lack of communications services on its lands, the Standing Rock Tribal government made the decision to establish its own mobile service provider and, in the process, broke ground for other Tribally owned companies.

While not every Tribal government will decide that self-provisioning is the chosen approach, Standing Rock illustrates what can be accomplished when that path is chosen.

In the context of universal service reform, Standing Rock's experience informed the Commission's decision to adopt certain provisions designed to facilitate self-provisioning and ETC designation. Only ETCs are eligible to participate in the Mobility Fund and Tribal Mobility Fund.[105] For Tribally-owned applicants only, however, the Commission adopted rules allowing those applicants to participate in the auctions as long as their ETC applications were submitted by the time the short form auction applications were due.[106] This provision affords Tribal governments the opportunity to seek ETC designation and to participate in the auctions at the same time. Standing Rock, and the nine other Tribally owned ETCs, paved the way for this opportunity.

This is not to say that the path to becoming an ETC, as well as a viable and successful business, is an easy one. Particularly on the remote and sparsely populated Standing Rock Reservation, the challenges are many. One critical issue is funding. As a successful bidder in Auction 901, Standing Rock will have the opportunity to use Mobility Fund support to expand the critical services it is providing on the Reservation – the very essence of universal service.

Access to Spectrum for Tribal Nations

Another issue of critical importance to Standing Rock and future Tribally owned wireless ETCs is access to spectrum. In an effort to address the limited availability of wireless services on Tribal lands, the Commission launched the *Spectrum Over Tribal Lands NPRM* in 2011, seeking comment, among other things, on proposals to create new opportunities for Tribes to gain access to spectrum. The Commission sought comment on those proposals, including expansion of the Tribal priority, secondary market processes, and a build-or-divest process, with the objective of promoting greater use of spectrum over Tribal lands. Tribal governments and their associations are in agreement that greater access to the spectrum over their lands for Tribal Nations is a linchpin to bringing Tribes and the wireless industry together to develop sustainable commercial wireless networks on Tribal lands. Tribal Nations and their inter-Tribal government associations offer continued support for consultation sessions and industry meetings to develop a record in support of these proposals.[107] The *Spectrum Over Tribal Lands NPRM* continues to be a major consultation priority for ONAP in 2013, so that other Tribally owned companies can follow in Standing Rock's footsteps, and more meaningful deployments can flourish between the wireless industry and Indian Country.

The Critical Nature of Lifeline and Link Up on Tribal Lands

The continued viability of the Tribal Lands Lifeline program, as well as the preservation of the Tribal Lands Link Up program, is also of critical importance to ETCs serving Tribal lands and has become integral to the business models and deployment plans of many of these companies, perhaps best exemplified by Tribally owned and operated ETCs like Standing Rock Telecom.[108] For many Tribal ETCs – Standing Rock Telecom in particular – even the names of the Lifeline and Link Up programs resonate more soundly, given the very high levels of unemployment they face, the very high percentage of families with incomes well under the Federal Poverty Guidelines, and the remote nature of the Reservations. The mobile service that Standing Rock provides at a discount to its low-income consumers literally provides a lifeline in the form of access in the event of an emergency, as well as a link or conduit for education, health care, economic or job opportunity, and basic communication.

ONAP is the face of the Commission in Indian Country, and in few places is that more apparent than Standing Rock. Since its creation in 2010, ONAP has forged a strong government-to-government relationship with the Standing Rock Tribal government and its Tribally owned ETC, Standing Rock Telecom. The building of this relationship was strengthened by personal contact when members of the ONAP staff visited the Standing Rock Reservation in 2011, and further solidified by frequent communication. It is critically important that ONAP have the opportunity to continue building on that relationship, which is accomplished most powerfully with face-to-face contact.

ONAP's 2012 Broadband and Telecom 101 Training Sessions

May 9-11: Washington State on the Swinomish Indian Tribal Community's Reservation, held in coordination with the Affiliated Tribes of Northwest Indians

May 23-24: Southern California on the Pala Indian Reservation, held in coordination with the Southern California Tribal Chairmen's Association

July 27: Anchorage, AK, hosted by ONAP at the invitation of U.S. Senator Mark Begich, who participated in a panel and provided a keynote address

July 31: Fairbanks, AK, held in coordination with the Tanana Chiefs Conference at their longhouse meeting room facility

September 12-13: Wisconsin on the Reservation of the Lac du Flambeau Band of Lake Superior Chippewa Indians, and held in coordination with the Great Lakes Inter-Tribal Council

September 18-19: Oklahoma, held for Oklahoma and Southern Plains Tribes at the Gaylord School of Communications and Journalism on the campus of the University of Oklahoma

A New Approach to Training in 2012

ONAP's "Tribal Broadband and Telecom 101" Events

In 2012, ONAP initiated a new Tribal training and consultation program, developed to be more responsive to the needs and requests of Indian Country. The decision to move toward a more targeted regional approach with smaller but more engaged and vested audiences was intended to create a far more interactive approach than in the past, as well as to provide opportunities for meetings with individual Tribes. It is with these goals in mind that *Tribal Broadband and Telecom 101*events were conceived, in coordination with Tribal leaders nationwide. The name itself, with the "101" designation, so common a concept reflecting a course of study and intellectual discourse, was itself a suggestions of the members of the FCC-Native Nations Broadband Task Force. Indeed, in many events, Tribal leader Task Force members played key roles in planning and hosting their respective training and consultation events.

ONAP conducted six *Tribal Broadband and Telecom Training 101* programs from May through September 2012 in venues across the country.

ONAP's training is aimed primarily at Tribal leaders, representatives, managers, and planners interested either in providing broadband and other communications services themselves, partnering with other entities (including carriers), or working with existing carriers. ONAP's training also aims to meaningfully involve the carriers or providers that serve Tribal lands, in an effort to provide the platform for their engagement with Tribal Nations and the Commission alike. A principal goal of the *Tribal Broadband and Telecom 101* events included informing Tribes about Commission policy initiatives affecting Tribal lands and soliciting Tribal input into Commission dockets. Expert staff from across the Commission participated in conducting the training workshops. Training sessions also included specific set-aside time and space to engage and consult individually with Tribal Nation representatives about general or specific issues and challenges they face.

ONAP worked carefully and deliberately, in detailed and highly cost-sensitive budget and program planning, to select training consultation venues that also responded to high levels of interest expressed by Tribal Nations or inter-Tribal government organizations, and to reach Tribal Nations in regions of the country not previously visited by the Commission. While training programs were regionally focused and the specific interests of Tribes in regions where trainings were held were emphasized, certain elements of the agendas were similar and Tribal leaders and representatives from all parts of Indian Country were welcome and invited to attend any of the training sessions. For example, while on the Pala Indian Reservation in California, Tribal representatives came from as far away as Alaska and Wisconsin. In Wisconsin, attendees came from South Dakota and, in Oklahoma, Tribal representatives from as far away as Florida and Maine attended.

ONAP's Tribal Broadband and Telecommunications 101 Workshop in Anchorage, AK, included a keynote address by U.S. Senator Mark Begich (left) who also participated in a panel. Senator Begich is joined by ONAP Chief Geoffrey Blackwell and Deputy Chief Irene Flannery.

ONAP's design of the training format was and is intentionally flexible. On multiple occasions in 2012, agendas were adjusted to address more fully issues of particular interest to a region or a set of attendees. ONAP coordinated with Tribal leaders smoothly and in real time to accommodate training and consultation needs. At the training on the Pala Indian Reservation in California, for example, an afternoon presentation session was shortened when a majority of the participants expressed interest in actually visiting in person the nearby network facilities of the Tribal Digital Village. As discussed more fully above, this program of the Southern California Tribal Chairmen's Association is bringing broadband to 17 Reservations using a wireless network of point-to-point and point-to-multipoint links covering some 350 miles. Visiting the facility afforded participants an inside perspective on how the Tribal Digital Village Network is providing broadband solutions for Tribes in the region and powerfully illustrated, in a real world setting, many of the topics addressed in the training sessions. At the same training session, a separate visit was made to KOPA - 91.3 FM Pala Rez Radio, the nearby radio station owned and operated by the Pala Band of Mission Indians.

It is important to reiterate and further explain that centrally important to the success of these training sessions was the partnership with, and participation by, Tribal leader members of the FCC-Native Nations Broadband Task Force.

FCC-Native Nations Task Force Members were instrumental in conducting regional programs in 2012 and were invaluable partners. These Task Force members, pictured above from left to right, and the events they supported are: Matthew R. Rantanen, Director of Technology, Tribal Digital Village, Pala Band of Mission Indians, training conducted on the Pala Reservation in CA; Pearl Mikulski, Community Services Vice President, Kawerak, Inc., ONAP meetings held in Nome, AK; James Williams, Chief Information Officer, Tanana Chiefs Conference, ONAP meetings and training conducted in Ruby and Fairbanks, AK; Michael J. White, Director of Information Technology and Records Management, Iowa Tribe of Oklahoma, training conducted in Oklahoma; Honorable Jeffrey Harjo, Council Member, Seminole Nation of Oklahoma, training conducted in Oklahoma; and (not pictured) Debby Gallenberg, IT Manager, Mole Lake Band of Lake Superior Chippewa, Sokaogon Chippewa Community, training conducted on the Lac du Flambeau Reservation in WI.

Task Force members were instrumental in planning the regional programs, arranging for venues, participating in panel sessions, identifying Tribal speakers, handling registrations, and providing support staff. Without the invaluable guidance and assistance of the Task Force members, the workshops would not have been nearly as successful as they were. In certain venues, such as the Swinomish Reservation in Washington State, and in Oklahoma, Task Force members arranged additional meetings with other Tribal leaders, to make introductions and discuss shared priorities in more detail. For any federal agency, this type of welcoming involvement by insiders to Indian Country, and treatment as an insider, is a rare occurrence and always beneficial to the federal agency's work. In the case of ONAP and the Commission, it is transformative to our policies and progress on Tribal government matters.

Training Sessions

Training agendas covered a vast array of topics presented and discussed at multiple levels of importance and depth of detail. The most basic level included an overview of the Commission and ONAP, as well as the Tribal policy agenda that the Commission initiated on Native Nations Day – March 3, 2011. Panels also addressed these and derivative policy initiatives affecting Tribal lands, including universal service and intercarrier compensation reform, the Mobility and Tribal Mobility Funds, Tribal Lands Lifeline and Link Up, and new rules creating opportunities to set up Tribally-owned and operated radio stations. Panels also addressed opportunities to comment on and meaningfully participate in Commission proceedings, such the *Spectrum over Tribal Lands NPRM*, as well as nuts-and-bolts topics such as how a Tribal entity can become an ETC. Experts from the Media Bureau, WCB, and WTB were important partners and presenters in training session panels. While the formal agendas were similar in many ways, the focus of each training was infused with regional concerns and issues. For example, radio broadcast issues were of primary interest in Wisconsin, while ETC designation and universal service issues were of most interest in Alaska and Oklahoma.

A favorite among Commission participants, Tribal attendees, and industry representatives was the Tribal leaders' panel, in which Tribal leaders, representatives, and IT or telecom managers candidly shared their experiences with communications challenges they encountered and how they identified or developed solutions to address them. This was useful not only for Tribal training attendees, but for all involved in gaining a better understanding of Tribal communications issues and concerns. This type of "voices from the trenches" experience will be invaluable in ONAP's plans for future training sessions and as it formulates subsequent Tribal policy initiatives with Commission Bureaus and Offices.

For the first time ever in the history of the Commission's Tribal training work, ONAP built time into the agendas, and budgeted for separate private meeting spaces, for one-on-one meetings with Tribal leaders to address Tribal-specific issues or to clarify or expand on information presented in sessions. For instance, in Wisconsin, individual meetings were held with leaders and representatives of six separate Tribal Nations to discuss a wide range of issues, from general concerns about broadband and telecommunications challenges and priorities, to obtaining ETC designation, protecting sacred sites and Tribal cemeteries in the face of deployment involving trenching, increasing Lifeline subscribership, and petitioning the Commission for an FM radio allotment.

There have been several important outcomes resulting from the *Tribal Broadband and Telecom Training 101* training and consultation events held in 2012. For example, there is increased Tribal government participation in Commission proceedings in the form of comments filed and individual consultations on issues such as universal service reform, ETC designation, and access to spectrum over Tribal lands. Training sessions on the radio Tribal Priority have resulted in several Tribal governments exploring with ONAP and the Audio Division of the Media Bureau options for starting their own radio stations. Panels on universal service reform and the Mobility and Tribal Mobility Funds have led to the submission of ETC designation petitions from Tribal governments. Based on levels of contact and inquiries, many more petitions are expected. And following the training sessions held in conjunction with the Affiliated Tribes of Northwest Indians in Washington State, Tribal IT and telecom managers formed a new regional working group to share information on common challenges, work with ONAP and the Commission, and learn from others' experiences in developing proposals to provide communications services for their Tribes. All of these outcomes were made possible by *Tribal Broadband and Telecom 101* training seminars that brought together Tribal leaders and Tribal representatives, industries that serve Tribal lands, and the Commission, all of whom shared experiences and learned from one other.

Native Learning Lab

In 2012, ONAP introduced the Native Learning Lab, which uses an arrangement of special low-cost, scaled down but still state of the art laptops, wireless routers, and printers, all set up in a separate meeting room at training venues and used to acquaint participants with the Commission's web-based resource systems and applications. The Native Learning Lab provides interactive, computer-based online tutorials to help participants understand the Commission's auctions process, use Commission mapping tools, identify spectrum licenses on Tribal lands, pinpoint areas that may be eligible for support under the new Mobility and Tribal Mobility Funds, learn how to access information on U.S. Department of Agriculture programs, and learn about how to interact with the Commission on regulatory proposals. Virtually any relevant potential instruction or learning opportunity that can occur through the use of these technological tools and the availability of the Internet occurs in the Native Learning Lab. It is the physical embodiment of the goals and principles of the Commission's Tribal Policy Statement, envisioning the use of technology and creative application of the Commission's intellectual resources in consulting with Tribal Nations.

One-on-one and small group Native Learning Lab instruction was conducted with the invaluable and ever-enthusiastic on-site support of personnel from the Office of Managing Director's Information Technology Center. Training and instruction at the Learning Lab has resulted in the submission of comments from new participants in Commission dockets and rulemakings. By demystifying Commission processes and systems, and through a hands-on introduction to them, Tribal leaders and representatives indicated that they could now approach the Commission with new confidence and enthusiasm. In these intimate sessions, Commission staff made accessible and relevant the systems that previously intimidated, and an important level of actual broadband adoption occurred among the communications "movers and shakers" at their respective Tribal Nations. Equally important, Tribal leaders and managers have developed working relationships with Commission staff in the Native Learning Lab and at training sessions and are now comfortable contacting staff later when questions arise or when additional information or guidance is needed. Commission instructors and assistants simply lost count of how many "A-Ha!" moments occurred in the Native Learning Lab with their Tribal Nation counterparts.

Reservation Visits

Visits to other nearby or regional Reservations and Tribal government offices and institutions were often conducted in conjunction with *Tribal Broadband and Telecom 101*training sessions. These visits enabled ONAP to maximize the travel dollars expended and empowered Tribal leaders by allowing Commission staff to meet with them on their lands and witness first-hand the impact of inadequate communications infrastructure and services. In 2012, ONAP staff visited a number of Reservations and Tribal communities in Oklahoma, Washington, Wisconsin, New Mexico, Arizona, Maine, Rhode Island, California, Nebraska, and Native Villages in Alaska that the Commission had never before visited. Meetings and consultations covered such topics as tower siting, obtaining a broadcast license, getting ETC designation for a Tribally owned entity, working with carriers under the Tribal government engagement obligation, helping Tribes evaluate the best approaches for the provision of communications services to their Tribal lands, and myriad other topics and issues. Visits to Tribal radio stations, health care facilities, schools, sacred traditional cultural properties, community centers, businesses, and other Tribal community anchor institutions gave ONAP and other Commission representatives the information necessary to recommend to the Commission solutions for specific, real world problems faced by Tribal Nations. These insightful visits provided information and perspectives that can be only developed in the field, and will help ONAP and the entire Commission develop the agenda for future policy initiatives and better coordination among Bureaus and Offices. This meaningful, on-the-ground Tribal consultation will also increase the quality of Tribal policy initiatives at the Commission, in terms of the ability of those initiatives to have definable impact on the persistent digital divide on Tribal lands. Maintaining an actual presence in Indian Country, and working directly with Tribal governments, measurably helps the Commission move the needle on services to Tribal lands.

In keeping with the spirit of the nation-to-nation relationship, certain Reservation meetings were scheduled on an ad hoc basis. For example, Tribal participants at the Oklahoma training in September requested a continuation of the *Tribal Broadband and Telecom Training 101* after the conclusion of the formal program. On the day after the training, ONAP staff met with telecommunications and IT professionals from five Tribal Nations in the administrative offices of the Sac and Fox Nation. Topics included challenges with the Tribal Lands Lifeline and Link Up programs, concerns about the presence of certain low-cost wireless providers, and the need for more effective outreach. This meeting, and others like it, were the natural and important progression of the Commission's training program; as the "communications IQ" of several Tribal representatives grew, issues associated with the availability and quality of broadband services on Tribal lands in Oklahoma and the broadcast opportunities available through the Tribal Priority took center stage. Following the Washington State training program, the ONAP team drove hundreds of miles to meet with Tribal Nations on their lands and to discuss and observe first-hand the communications challenges they face. ONAP visited the Coeur d'Alene Reservation in Idaho, which is a showcase for the successful provision of broadband services to Tribal entities, businesses, and members. ONAP staff also visited two Tribal governments on the opposite side of Washington State from the Swinomish Reservation – the Confederated Tribes of the Colville Reservation, which faces serious connectivity challenges over a large geographical area; and the Spokane Tribe. Riding a barge across the Columbia River, this was the Commission's first visit to the Spokane Tribe's Reservation and it gave ONAP an opportunity to meet with key Spokane leaders on the Tribe's communications needs – leaders not otherwise able to meet with ONAP, as no budget was available for them to come to the Commission's headquarters in Washington, DC. ONAP-conducted training and Tribal consultation sessions will continue to evolve in 2013 and beyond. Several more *Tribal Broadband and Telecom 101* programs are planned, so that more Tribal governments with limited experience in broadband and telecommunications can learn the basics and become fully versed, fluent, and engaged in the Commission's Tribal Nation policy initiatives and regulatory processes. ONAP also has plans to work closely with Tribal governments, inter-Tribal government organizations, and the FCC-Native Nations Broadband Task Force to determine what the next level of training – pcrhaps *Tribal Broadband and Telecom 201 or 301* – should encompass.

ONAP visited the Spokane Indian Reservation in eastern Washington state for the first time in 201, and met with economic development officials of the Spokane Tribe. This required a ferry ride across Roosevelt Lake, part of the Columbia River.

ACKNOWLEDGEMENT OF THE FCC-NATIVE NATIONS BROADBAND TASK FORCE

ONAP would like to acknowledge and express its heartfelt gratitude to the members of the FCC-Native Nations Broadband Task Force. Without the invaluable and substantive input and contributions from the Task Force, the Commission's renewed regulatory agenda for Tribal Nations and its new approach to training would not have taken shape. It is important to note that each of these individuals has served without compensation of any kind from the Commission. Members are listed alphabetically within the appropriate categories.

Elected Tribal Leaders

- Honorable Marlin Fryberg: Tulalip Tribes
- Honorable Joe Garcia: Ohkay Owingeh Pueblo – Task Force Co-Chair
- Honorable Jeffrey Harjo: Seminole Nation of Oklahoma
- Honorable Bradley John: Seneca Nation of Indians
- Honorable Jim Shakespeare: Northern Arapaho Tribe

Appointed Tribal Leaders

- Mark Bilton-Smith: Cow Creek Band of Umpqua Tribe of Indians
- Lewis Christman: Tule River Indian Tribe
- Valerie Fast Horse: Coeur d'Alene Tribe
- Debby Gallenberg: Sokaogon Chippewa Community
- Pearl Mikulski: Kawerak
- Carroll Onsae: Hopi Tribe
- Matthew Rantanen: Pala Band of Mission Indians
- Brian Tagaban: Navajo Nation
- Michael White: Iowa Tribe of Oklahoma
- James Williams: Tanana Chiefs Conference

Senior Commission Staff

- Geoffrey Blackwell: Task Force Co-Chair
- Kirk Burgee: Wireline Competition Bureau
- Irene Flannery: Consumer and Governmental Affairs Bureau
- David Furth: Public Safety and Homeland Security Bureau
- Jane Jackson: Wireless Telecommunications Bureau
- Mark Lloyd: Office of General Counsel
- Kris Monteith: Media Bureau ***(position currently vacant)***
- Robert Nelson: International Bureau
- Jamison Prime: Office of Engineering and Technology
- Thomas Reed: Office of Communications Business Opportunities
- Suzanne Tetreault: Enforcement Bureau ***(position currently vacant)***

ACKNOWLEDGEMENT OF BUREAU AND OFFICE MANAGERS AND STAFF INVOLVED IN WORKING WITH TRIBAL NATIONS

A recurring theme throughout this Annual Report is one of teamwork and partnership. ONAP would not have been able to accomplish all that it did in 2012 without the invaluable contributions of colleagues from across the Commission. We offer our sincere gratitude to everyone with whom ONAP has worked so closely, and we single out the following individuals for particular recognition, listed alphabetically within their respective Bureaus:

Consumer and Governmental Affairs Bureau

- Esther Butler
  Administrative Management Specialist
- Tracy Epps-Jones
  Administration Management Specialist
- Tamika Jackson
  Associate Bureau Chief for Management
- Kris Monteith
  Acting Chief
- Toni Simmons
  Administrative Management Specialist

Media Bureau

- James Bradshaw
  Deputy Chief, Audio Division
- Rudy Bonacci
  Electronics Engineer, Audio Division
- Peter Doyle
  Chief, Audio Division
- Thomas Nessinger
  Senior Counsel, Audio Division
- Lisa Scanlan
  Assistant Chief, Audio Division

Office of the Managing Director

- Kim Bassett
  Budget Officer

- Nelson Davis
  Customer Service Representative
- Carolyn Dickey
  Budget Analyst
- Earl Ivey
  Carlson-Wagonlit
- David Robbins
  Managing Director
- Mark Stephens
  Chief Financial Officer
- Jacqueline Walker-Robinson
  Financial Management Specialist
- Darshan "Doc" Williams
  IT Specialist, Information Technology Center
- Tenecia Williams
  Financial Systems Specialist

Wireless Telecommunications Bureau

- Stephen DelSordo
  Federal Historic Preservation Officer, Wireless Telecommunications Bureau
- Sue McNeil
  Special Counsel, Auctions and Spectrum Access Division
- Jeffrey Steinberg
  Deputy Chief, Spectrum and Competition Policy Division
- Dorothy Stifflemire
  Outreach Marketing Specialist, Wireless Telecommunications Bureau
- Margaret Wiener
  Chief, Auctions and Spectrum Access Division
- Anne Marie Wypijewski
  Attorney Advisor, Spectrum and Competition Policy Division

Wireline Competition Bureau

- Joseph Cavender
  Assistant Chief, Telecommunications Access Policy Division
- Patrick Halley
  Legal Advisor, Office of the Bureau Chief (now Deputy Director, Office of Legislative Affairs)
- Jonathan Lechter
  Attorney Advisor, Telecommunications Access Policy Division

CONCLUSION

Community-oriented and truly effective deployment of communications technologies within Indian Country has the potential to level the negative social, cultural, and economic impacts that history has caused Tribal Nations and Native communities to endure. New commercial, educational, and health care opportunities, as well as social stability and quality of life issues, may genuinely be addressed though broadband and other communications mediums. Working towards these ends, the Commission's coordination, engagement, and training with Tribal Nations experienced unprecedented levels in fiscal year 2012. The Commission continued to build upon its reputation among Tribal Nations as one of the most active, creative, and determined of federal agencies in solving the often recalcitrant problems of the digital divide in Indian Country.

With the creation of ONAP, the Commission has adopted a reinvigorated regulatory agenda for Tribal lands nationwide, involving critical elements of consultation, outreach, and training. This renewed effort is in major motion, and must be maintained, as there is much more good work to be done, with additional milestones and initiatives to be achieved. The rules found throughout the regulations of the Commission regarding Tribal lands are becoming more and more interrelated with the comprehensive solutions to the lack of services on Tribal lands. Tribal lands spectrum licensing, Mobility and Tribal Mobility Fund reverse auctions, Lifeline assistance, Link Up support, the Commission's authority to designate ETCs on Tribal lands, and the Commission's renewed consultation, training, and engagement programs all play a role in the development of genuine solutions for Tribal Nations.

Through ONAP's work across the entire FCC, the Commission has identified key strategic indicators for success in addressing the lack of broadband and other critical communications technologies among American Indian Tribes, Alaska Native Villages, and Native communities, such as Hawaiian Home Lands. These important indicators of success are reflected in the Commission's new rules, proposed rules, and new policies with respect to Tribal Nations. These indicators of success include new levels of dialogue and reporting, new licensing priority opportunities, and increased support and investment through universal service support mechanisms.

New and important solution-oriented dialogue goals are found both in the renewed nation-to-nation consultation between the Commission and Indian Country, and in rules for a new "meaningful engagement" between universal service-supported carriers and the Tribal Nations whose lands they serve. The reporting requirement of the Tribal government engagement obligation will provide a basis for Tribal Nations, the Commission, and the supported carriers to address future needs and opportunities. It has the potential to become the detailed solution roadmap to the digital divide on a reservation by reservation, Tribal Nation by Tribal Nation basis.

The new radio broadcast licensing Tribal priority and the Commission's commercial mobile radio spectrum rulemaking proposals for Tribal lands recognize a fundamental component to finding a solution in the "one-size-fits-none" Indian Country paradigm. That is, very often the individual Tribal Nation itself must become invested and involved in the solution or delivery of services. The need for spectrum-based services is pervasive and, in many places across Indian Country, the lack of wireless telephony and broadband is a matter of life or death. As Tribal Nations begin to take their place at the table of broadcast media, the prevalent misperceptions of Tribal Nations will be corrected, and Tribal cultural values will be preserved at the same time as public safety is protected through the long adopted broadcast radio services.

Several of the reasons why any Tribal Nation governs its people and its lands – preserving and advancing Tribal Nation community culture and providing future opportunity for its people – are the same reasons why it must become involved itself in identifying and finding solutions to the lack of services. When a Tribal Nation steps up to work with the Commission and the communications industries, solutions follow.

The Commission, with ONAP and ONAP's coordination throughout the Bureaus and Offices of the FCC, has laid the groundwork for the path forward. Multiple important new regulatory priorities and opportunities are taking effect in fiscal year 2013, such as the Tribal Mobility Fund Phase I reverse auction. The Tribal consultation and coordination priorities of fiscal year 2013 – the Tribal Mobility Fund, Tribal government engagement obligation, *Spectrum Over Tribal Lands Notice of Proposed Rulemaking*, cultural preservation review of non-compliant towers, ETC designation petitions to serve Tribal lands, and the radio broadcast Tribal Priority – will provide the opportunity to maintain the forward momentum of the Commission's agenda for Tribal Nations and Native communities, and its critical mission to close the digital divide in Indian Country.

APPENDIX A: LIST OF 2012 TRAVEL, TRAINING, AND MEETINGS IN INDIAN COUNTRY

Commission-wide travel relating to Tribal Nations in 2012 included representatives from the Consumer and Governmental Affairs Bureau, Media Bureau, Office of Managing Director, Wireless Telecommunications Bureau, and Wireline Competition Bureau

First Quarter

01/08-13 -- Santa Fe, NM: FCC Native Nations Task Force Meeting – conducted meeting; consultations; Reservation visits/meetings

01/31-02-06 – Phoenix, AZ: Tribal Telecom Conference 2012 – presented, exhibited; consultation; visit/meeting on the Gila River Reservation

02/12-18 – Olympia, WA: Affiliated Tribes of Northwest Indians Conference – presented

02/26-03/02 – Las Vegas, NV: Reservation Economic Summit – presented, exhibited, consultation

03/19-21 – New Orleans, LA: Native American Finance Officers Association Conference – presented

Second Quarter

04/02-05 – Durant, OK: To Bridge A Gap Conference – presented, consultation

05/09-16 – Anacortes, WA: Regional Training – conducted Native Learning Lab, consultations; visits/meetings on the Colville and Spokane Reservations in Washington State and the Coeur d'Alene Reservation in Idaho

05/22 – Marksville, LA: United South and Eastern Tribes 2012 Semi-Annual Meeting – presented, consultation

05/21-26 – Pala, CA: Regional Training – conducted Native Learning Lab, consultation; Tribal Enterprise visits

05/30-06/02 – Arizona/New Mexico – AZ-NM Telecommunications Assn. Conference; consultation

Third Quarter

06/16-21 – Lincoln, NE: NCAI Mid-Year Conference and Task Force Meeting – presented, exhibited, conducted Native Learning Lab, consultation

07/11-15 – Albuquerque, NM: Native Public Media Summit – presented; consultations; Reservation meeting at Ohkay Owingeh (San Juan) Pueblo

07/22-08/01 – Alaska (Kotzebue, Noatak, Nome, Anchorage, Fairbanks, and Ruby): Regional Training; community meetings; consultations; visits with Alaska Native Village governments in Kotzebue, Noatak, Nome, and Ruby

08/27-31 – Rhode Island and Maine: Consultations, visits meetings on the Narragansett Indian Reservation in Rhode Island, and with the Houlton Band of Maliseet Indians and the Aroostook Band of Micmacs in Maine

09/10-12 – Suquamish, WA: Annual Training Conference of the National Association of Tribal Historic Preservation Officers – presented

09/10-14 – Lac du Flambeau, WI: Regional Training – conducted Native Learning Lab, consultation; Reservation visit

09/17-21 – Oklahoma City: Regional Training – conducted Native Learning Lab, consultation; meeting/visit to the Muscogee (Creek) Nation Reservation; meeting with telecommunications and IT professionals from five North Central Oklahoma Tribes on the Sac and Fox Nation

09/24-27 – Pendleton, OR: Annual conference of the Affiliated Tribes of Northwest Indians – presented, conducted Native Learning Lab, exhibited; Tribal enterprise visit

09/24-30 – Phoenix, AZ: Meetings of the Inter Tribal Council of Arizona and National Tribal Telecommunications Association – presented; consultation; visits/meetings at Gila River and Hopi Reservations

09/25-27 -- Devil's Tower, WY, and Medicine Butte, SD: Tower site visits

Fourth Quarter

10/21-26 – Sacramento, CA: NCAI Annual Conference – presented, consultation

APPENDIX B: 2012 HEADQUARTERS MEETINGS AND CALLS WITH TRIBAL LEADERS, CARRIERS, AND PARTIES INTERESTED IN THE PROVISION OF COMMUNICATIONS SERVICES IN INDIAN COUNTRY

Throughout the year, and together with many other Bureaus and Offices from across the Commission, ONAP held substantive meetings and conference calls with elected Tribal leaders and officials from Tribal governments and other entities, Tribally owned telecommunications carriers, and inter-Tribal organizations. ONAP, joined by other Commission officials, also met with a broad range of the leadership and representatives of carriers serving Tribal lands, as well as industry associations. These meetings often focused on open dockets and other matters before the Commission. With respect to Tribal meetings, they often followed up on discussions begun at field training programs, Tribal conferences, and during consultations and listening sessions.

Meeting topics included obtaining ETC certification, the transformation of USF and the impact on Tribes and carriers serving Tribal lands, changes to the Tribal Lifeline and Link Up programs, the Mobility Fund and Tribal Mobility Fund, the Tribal government engagement obligation, obtaining spectrum, starting or obtaining a Tribal radio or TV station, and filing comments and petitions relating to Commission rules and rulemakings affecting Tribal Nations, Native communities, and Tribal lands.

The list below is not nearly exhaustive, but is illustrative of the broad breadth of entities with which ONAP held substantive calls and meetings during 2012. Calls and meetings were conducted with:

January 2012

1/23 – Smith Bagley, Inc.

February 2012

2/15 – United South and Eastern Tribes, Inc.

March 2012

3/6 – Acoma Pueblo
3/6 – Karuk Tribe
3/6 – Warm Springs Telecommunications Company
3/7 – Indian Health Service
3/8 – Gila River Indian Community and
Gila River Telecommunications, Inc.
3/8 – United South and Eastern Tribes, Inc.
3/9 – Southern California Tribal Chairmen's Association
3/9 – National Tribal Telecommunications Association
3/15 – Affiliated Tribes of Northwest Indians
3/15 – Seneca Nation of Indians
3/19 – South Dakota RLECs
3/21 – Hopi Tribe
3/22 – Confederated Tribes of the Colville Reservation
3/28 – Mescalero Apache Telecom, Inc.

April 2012

4/9 – Red Lake Band of Chippewa Indians
4/19 – Tulalip Tribes
4/19 – Sacred Wind Communications
4/20 – Affiliated Tribes of Northwest Indians
4/25 – Southern California Tribal Chairmen's Association
4/26 – Inter Tribal Council of Arizona
4/27 – San Carlos Apache Tribe

May 2012

5/1 – Affiliated Tribes of Northwest Indians
5/2 – Hopi Tribe
5/4 – Affiliated Tribes of Northwest Indians
5/7 – Hopi Telecommunications, Inc.;
Gila River Telecommunications, Inc.; and Saddleback Communications
5/17 – Association of Tribal Archives, Libraries,
and Museums 5/18 – National Congress of American Indians
5/19 – Native Public Media

June 2012

6/7 – Mescalero Apache Tribe, San Carlos Apache Tribe, San Carlos Apache Telecommunications and Utilities, Inc., Mescalero Apache Telecommunication, Inc.
6/12 – Hopi Telecommunications, Inc.; Gila River Telecommunications, Inc.; and Saddleback Communications
6/13 – San Carlos Apache Telecommunications and Utilities, Inc. 6/26 – Eastern Band of Cherokee Indians

July 2012

7/5 – Ione Band of Miwok Indians
7/16 – Smith Bagley, Inc.
7/18 – Lower Brule Sioux Tribe
7/19 – Loneman School, Oglala, SD

August 2012

8/8 – Tanana Chiefs Conference
8/8 – Red Lake Band of Chippewa Indians
8/13 – Navajo Nation
8/13 – Seminole Tribe of Florida

8/15 – Tulalip Tribes
8/22 – Navajo Nation, Sacred Wind Communications, and Frontier Communications
8/24 – Nez Perce Tribe

September 2012
9/6 – Seminole Tribe of Florida
9/6 – John Staurulakis, Inc.
9/6 –Cherokee Nation
9/7 – National Tribal Telecommunications Association

October 2012
10/10 – Bad River Band of Lake Superior Chippewa
10/11 – Alaska Federation of Natives
10/18 – Navajo Tribal Utility Authority

November 2012
11/1 – PCIA-The Wireless Infrastructure Association
11/7 – Hualapai Tribe
11/7 – Mississippi Band of Choctaw Indians
11/15 – US Telecom, Inc.; CTIA-The Wireless Association; NTCA-The Rural Broadband Association; and several member companies
11/16 – National Cable & Telecommunications Association
11/26 – Advisory Council on Historic Preservation
11/28 – Loneman School, Oglala, SD
11/29 – Loneman School, Oglala, SD
11/29 – Mobi Wireless

December 2012
12/3 - Standing Rock Telecommunications, Inc.
12/6 - Mescalero Apache Telecom, Inc.
12/6 - Keweenaw Bay Indian Community
12/13 - Sandwich Isles Communications, Inc.
12/18 – U.S. Government Accountability Office

APPENDIX C: EXAMPLE LETTERS FROM TRIBAL LEADERS

EDA | Jamestown S'Klallam Tribe
Economic Development Authority

Received & inspected
MAY 29 2012
FCC Mail Room

May 22, 2012

Julius Genachowski, Chairman
Federal Communications Commission
445 12th Street SW
Washington, DC 20554

RE: Tribal Telecom and Broadband 101 – Sponsored by the FCC and ANTI
Swinomish Casino and Lodge, Washington
May 09, 2012 through May 11, 2012

Dear Chairman Genachowski,

A very important lesson learned early in my business career is that an organization is known by the people it keeps. It is with great pleasure that I write to you to say the people representing the FCC, Office of Native Affairs and Policy at the recent conference in the State of Washington are extraordinary. The content of the conference was thoughtfully arranged and presented. It was useful, and understandable. It was made even more valuable by the interagency cooperation with the USDA telecommunications office. Every step of the way Chief Geoffrey Blackwell was making certain no one was left behind in the information loop. Time was well spent when he recapped the truly important points of the presentation, and dedicated sufficient time for questions and answers. This conference was far from the 'usual'. I came away believing that what I heard was accurate, and most importantly that we have high level knowledgeable and helpful friends at the FCC, who are genuine, and concerned about our success.

It's obvious that Mr. Blackwell has formed a culture in the Office of Native Affairs and Policy that reflects his sincerity through each of the staff members present. This may sound over-the-top, but I have never in all my years of experience found a group of people so in touch with what they are doing, and how they are doing it. They are generating respect and confidence for your organization.

Thank you for taking the time to read this. Thank you more for the people you keep, they will make you proud.

Sincerely,

Douglas Sellon, Executive Director
Jamestown S'Klallam Tribe Economic Development Authority

CC: W. Ron Allen, Chairman/CEO
Jamestown S'Klallam Tribe

Geoffrey C. Blackwell, Chief
FCC Office of Native Affairs and Policy

JAMESTOWN S'KLALLAM EDA
257 Business Park Loop
Sequim, Washington 98382
office 360.683.2025
fax 360.683.9583

JKT Development, Inc.
JKT CONSTRUCTION
15405 SE 37th Ave Suite 110
Bellevue, Washington 98006
office 425.732.6805
fax 425.732.6819

JKT Development, Inc.
JAMESTOWN EXCAVATING
1033 Old Blyn Highway
Sequim, Washington 98382
office 360.683.4586
fax 360.582-0120

Jamestown S'Klallam EDA
CARLSBORG MINI STORAGE
292 Business Park Loop
Sequim, Washington 98382
office 360.681.3536
fax 360.683.9583

The Eastern Band of Cherokee Indians

The Honorable Michell Hicks, Principal Chief
The Honorable Larry Blythe, Vice-Chief

Received & Inspected
JUL 11 2012
FCC Mail Room

June 21, 2012

Mr. Geoffrey C. Blackwell
Chief
Office of Native Affairs and Policy
Federal Communications Commission
Washington, DC 20554

Dear Mr. Blackwell,

I was recently made aware of new important FCC Releases that could potentially have a significant positive impact on the Eastern Band of Cherokee Indians. The information related to the FCC releases was provided to me by Mr. Bob Long and Mr. Ray Rose. Mr. Long, whom you met at the recent NCAI gathering, is our Tribal Telecom Manager and Mr. Rose is the Chairman of our Broadband Enterprise.

The Eastern Band of Cherokee, like many other tribes, struggles to provide the same level of quality internet connectivity and voice service that is available in so many areas across the nation. I am proud to say that despite the struggle our tribe recognized the absolutely critical need to provide our members with access to the internet and took action to solve the problem. Approximately four years ago we appropriated 1.875 million dollars as a capital investment to plan and implement a fixed wireless system to provide middle and last mile high speed internet service to our members and surrounding areas. Once the construction of the system was started we established the Cherokee Broadband Enterprise with its Board of Advisors to oversee the completion and operation of the system. The enterprise is wholly owned by our tribe.

To date,the enterprise has constructed a network operations center, four towers and several relays to begin improving service. At this time we are providing service to approximately 170 customers consisting primarily of residential and small business customers and are experiencing growth each month. The more important number may be the 200 plus tribal members who have requested service which unfortunately is not yet available in their area.

88 Council House Loop • P.O. Box 455 • Cherokee, NC 28719
Telephone: (828) 497-2771 or 497-7000
Telefax: (828) 497-7007

With each passing day the enterprise works aggressively to enhance the system through additional relays and other technologies to reach the next layer of customers. We are eagerly awaiting delivery of technology that we have purchased that will allow us to use the UHF spectrum formerly used for television broadcasting. Once we receive the equipment we will conduct tests to determine the performance of the technology and make a decision as to whether or not broader deployment is feasible and cost effective.

In addition, our tribe either is a partner in or wholly owns considerable fiber assets that have a tremendous potential to impact the tribe and the state of North Carolina. Our Broadband Enterprise has sole oversight of a portion of this fiber and collaborates with other entities on the best use of the fibers that are a part of the partnership.

Of particular interest to me and our tribe is the language in the FCC Releases related to the Tribal Mobility Fund, the important proposed rulemaking to improve wireless communications for Native Nations, and the requirement for Eligible Telecommunications Carrier (ETC's) to undertake meaningful engagement with Tribal Governments.

It is my understanding that it is important for each tribe to designate a person or group of persons as our representative(s) to work with the FCC and possibly other agencies to carry out the communications and activities necessary to take advantage of the various FCC programs and the benefits afforded by the approved and pending reforms.

Therefore; please be advised that the Chairman of the Cherokee Broadband Board of Advisors is designated as the representative to the FCC on behalf of the Eastern Band of Cherokee Indians to conduct business for the purpose of advancing and optimizing high speed wireless internet and voice services on and near our tribal lands. At this time Mr. Rose is our Chairman. He will work in conjunction with his fellow board members under their guiding documents to achieve the above goals.

In closing please allow me to express my appreciation for your efforts to create a level playing field for tribes and to help simplify the often complex matters related to providing broadband access to tribal members. It would be our pleasure to host you as our guest should you decide to visit Cherokee.

Sincerely,

Michell Hicks

Michell Hicks
Principal Chief
Eastern Band of Cherokee Indians

Sac and Fox Nation

920883 S. Hwy. 99 Bldg. A • Stroud, OK 74079

Principal Chief GEORGE THURMAN
Second Chief ORVENA (TWIGGY) GREGORY
Secretary JACKLYN K. WILLIAMS
Treasurer CARLA REED
Committee Member STELLA NULLAKE

Received & Inspected
OCT - 4 2012
FCC Mail Room

September 27, 2012

Federal Communications Commission
445 12th Street, SW
Washington, DC 20554

Re: 2013 Budget for the Office of Native Affair and Policy (ONAP)

To Whom It May Concern:

The Sac and Fox Nation would like to thank the Commission for its efforts during the last two years to correct the digital divide facing Indian Country. The impact of the Office of Native Affairs and Policy (ONAP) and the Native Nations Broadband Taskforce cannot be understated. Recent changes to Lifeline, auction requirements for FM frequencies, and set-asides in the Connect America Fund will go a long way toward bringing contemporary communications capabilities to our lands.

There is more work to be done. Last week ONAP held a training session here in Oklahoma. During the two day event and in the day that followed, FCC staff were able to hear first-hand from tribal staff and leaders regarding current problems, carrier abuse, and growing frustration surrounding the lack of technology that plague Indian Country each day.

We recognize that ONAP requires a larger than normal travel budget given the remoteness of most tribal headquarters. Their presence here, however, is critical to their understanding of current conditions and allows them to hear from individuals who are unable to travel to Washington or provide formal correspondence that adequately describes our plight.

Thank you again for your continued support of their Office and its mission. Natives work with others based upon relationships. ONAP has made great strides in repairing the FCC's relationship with tribes and we look forward to more collaboration and consultation during the upcoming year.

Sincerely,

George Thurman
Principal Chief
Sac and Fox Nation

Administration (918) 968-3526 Fax (918) 968-4837 □ Office of Government (918) 968-1141 Fax (918) 968-1142

Appendix D: Example Resolutions from Tribal Nations

NATIONAL CONGRESS OF AMERICAN INDIANS

The National Congress of American Indians
Resolution #SAC-12-019

TITLE: Request that the Federal Communications Commission Uphold the Tribal Government Engagement Obligation Provisions for Telecommunications Providers Serving Tribal Lands

WHEREAS, we, the members of the National Congress of American Indians of the United States, invoking the divine blessing of the Creator upon our efforts and purposes, in order to preserve for ourselves and our descendants the inherent sovereign rights of our Indian nations, rights secured under Indian treaties and agreements with the United States, and all other rights and benefits to which we are entitled under the laws and Constitution of the United States, to enlighten the public toward a better understanding of the Indian people, to preserve Indian cultural values, and otherwise promote the health, safety and welfare of the Indian people, do hereby establish and submit the following resolution; and

WHEREAS, the National Congress of American Indians (NCAI) was established in 1944 and is the oldest and largest national organization of American Indian and Alaska Native tribal governments; and

WHEREAS, on October 27, 2011, the Federal Communications Commission (FCC) adopted the *Universal Service Fund/Intercarrier Compensation Transformation Order*, in which the FCC concluded that eligible telecommunications carriers serving Tribal Lands and Tribal Governments should hold discussions regarding (1) a needs assessment and deployment planning with a focus on tribal community anchor institutions; (2) feasibility and sustainability planning; (3) marketing services in a culturally sensitive manner; (4) rights of way processes, land use permitting, facilities siting, environmental and cultural preservation review processes; and (5) compliance with tribal business and licensing requirements (collectively, the Tribal Government Engagement Obligation Provisions); and

WHEREAS, the FCC further delegated to the FCC's Office of Native Affairs and Policy (FCC-ONAP), in coordination with the Wireline Competition Bureau and the Wireless Telecommunications Bureau, the authority to develop specific best practices regarding the Tribal Government Engagement Obligation Provisions; and

WHEREAS, on July 19, 2012, FCC-ONAP released a *Further Guidance* developing specific procedures regarding the Tribal Engagement Obligations; and

WHEREAS, the United States Telecom Association (USTA) filed a petition with the FCC on August 20, 2012 requesting reconsideration and clarification of the *Further Guidance*; and

NCAI HEADQUARTERS
1516 P Street, N.W.
Washington, DC 20005
202.466.7767
202.466.7797 fax
www.ncai.org

WHEREAS, the NCAI, Native Public Media (NPM), the Gila River Indian Community, and the Gila River Telecommunications, Inc. – a wholly-owned entity of the Gila River Indian Community – filed replies in opposition to the USTA Petition reaffirming the critical importance of applying the Tribal Government Engagement Obligation Provisions to all eligible telecommunications carriers receiving and seeking Universal Service Fund support to all Tribal Lands; and

WHEREAS, the procedures established by the *Further Guidance* are timely due to the critical need to improve communications services on Tribal Lands immediately; and

WHEREAS, the Tribal Government Engagement Obligation Provisions and the Federal Communications Commission's *Further Guidance* by *Public Notice* are constitutionally sound because such procedures are intended to address the lack of communications services on Tribal Lands; and

WHEREAS, the oppositions filed by the aforementioned organizations and tribes further stated that the procedures established by the FCC's Connect America Fund Order and clarified by the *Further Guidance* are not unduly burdensome due to the numerous benefits that such procedures will offer to both Tribal Governments and eligible telecommunications carriers (ETCs) serving Tribal Lands.

NOW THEREFORE BE IT RESOLVED, that the NCAI hereby supports the procedures established in the FCC's Connect America Fund Order and clarified by the *Further Guidance* in that such procedures will provide benefits to both Tribal Governments and ETCs serving Tribal Lands; and

BE IT FURTHER RESOLVED, that NCAI calls for the FCC to immediately deny any current and future Petitions for Reconsideration and/or Clarification that would rescind the Tribal Government Engagement Obligation Provisions for ETCs and undermine the deployment of communications services on Tribal Lands; and

BE IT FINALLY RESOLVED, that this resolution shall be the policy of NCAI until it is withdrawn or modified by subsequent resolution.

CERTIFICATION

The foregoing resolution was adopted by the General Assembly at the 2012 Annual Session of the National Congress of American Indians, held at the Sacramento Convention Center from October 21-26, 2012 in Sacramento, California, with a quorum present.

Jefferson Keel
President

ATTEST:

Edward K. Thomas
Recording Secretary

NATIONAL CONGRESS OF AMERICAN INDIANS

The National Congress of American Indians
Resolution #SAC-12-021

TITLE: In Support of Universal Broadcast Service Over Tribal Lands

WHEREAS, we, the members of the National Congress of American Indians of the United States, invoking the divine blessing of the Creator upon our efforts and purposes, in order to preserve for ourselves and our descendants the inherent sovereign rights of our Indian nations, rights secured under Indian treaties and agreements with the United States, and all other rights and benefits to which we are entitled under the laws and Constitution of the United States, to enlighten the public toward a better understanding of the Indian people, to preserve Indian cultural values, and otherwise promote the health, safety and welfare of the Indian people, do hereby establish and submit the following resolution; and

WHEREAS, the National Congress of American Indians (NCAI) was established in 1944 and is the oldest and largest national organization of American Indian and Alaska Native tribal governments; and

WHEREAS, the U.S. Congress through the Public Broadcasting Act of 1967, as amended, declared "it is in the public interest to encourage the growth and development of public radio and television broadcasting, including the use of such media for instructional, educational, and cultural purposes;" and

WHEREAS, fifty-three Native licensed radio stations currently serve Tribal Communities across the United States providing critical news, information and Native language programming to over half a million listeners on the terrestrial network alone; and

WHEREAS, the U.S. Congress according to the 1967 Act appropriates funding on an annual basis for public radio and television through the Corporation for Public Broadcasting (CPB) "to complement, assist, and support a national policy that will most effectively make public telecommunications services available to all citizens of the United States;" and

WHEREAS, thirty of the fifty-three Native licensed non-commercial educational radio stations are in the CPB's Community Service Grant program receiving over $4 million in critical annual funding support; and

WHEREAS, the proposed sequestration of federal funding at 8.2% will result in a reduction of CPB's funding of $36 million for both radio and television stations and a reduction of 8.2% overall for Native radio stations who rely on CPB funding for their annual operations; and

NCAI HEADQUARTERS
1516 P Street, N.W.
Washington, DC 20005
202.466.7767
202.466.7797 fax
www.ncai.org

WHEREAS, the Native radio stations serve the most invisible, unserved and underserved populations in the United States and are among the most vulnerable to funding cuts due to the economic, rural and challenging nature of the tribal communities Native radio serves; and

WHEREAS, the Native radio stations are often the only means of news and information for tribal communities located in some of the most rural and geographically isolated regions of the United States.

NOW THEREFORE BE IT RESOLVED, that the NCAI urges Congress to support the CPB in providing a funding safety net for Native radio stations to ensure that these stations will remain on air to provide vital and critical programming including public safety communications for tribal communities; and

BE IT FURTHER RESOLVED, that the CPB engage in meaningful consultation with Tribal Nations on the future of public broadcasting that promotes greater universal service of valued public service media to current and more tribal communities; and

BE IT FINALLY RESOLVED, that this resolution shall be the policy of NCAI until it is withdrawn or modified by subsequent resolution.

CERTIFICATION

The foregoing resolution was adopted by the General Assembly at the 2012 Annual Session of the National Congress of American Indians, held at the Sacramento Convention Center from October 21-26, 2012 in Sacramento, California, with a quorum present.

President

ATTEST:

Recording Secretary

NATIONAL CONGRESS OF AMERICAN INDIANS

The National Congress of American Indians
Resolution #SAC-12-033

TITLE: In Support of a Dedicated Annual Budget for the Federal Communications Commission, Office of Native Affairs and Policy

WHEREAS, we, the members of the National Congress of American Indians of the United States, invoking the divine blessing of the Creator upon our efforts and purposes, in order to preserve for ourselves and our descendants the inherent sovereign rights of our Indian nations, rights secured under Indian treaties and agreements with the United States, and all other rights and benefits to which we are entitled under the laws and Constitution of the United States, to enlighten the public toward a better understanding of the Indian people, to preserve Indian cultural values, and otherwise promote the health, safety and welfare of the Indian people, do hereby establish and submit the following resolution; and

WHEREAS, the National Congress of American Indians (NCAI) was established in 1944 and is the oldest and largest national organization of American Indian and Alaska Native tribal governments; and

WHEREAS, the Federal Communications Commission (FCC) established the Office of Native Affairs and Policy (FCC-ONAP) to meaningfully engage with Tribal Nations on a government-to-government basis to address the dire nature and deep void of communications services on Tribal Lands; and

WHEREAS, since its establishment, the FCC-ONAP has consulted with Tribal Nations across the United States resulting in the most comprehensive engagement of Tribal Nations on the deployment and adoption of communications services on Tribal Lands; and

WHEREAS, the tribal engagement and consultation by the FCC-ONAP has resulted in rules that include Tribal Nations in the National Broadband Plan, the Connect America Fund and Tribal Mobility Fund, reforms to the Lifeline and Link Up Programs, other universal service reforms, and further rulemakings; and

WHEREAS, in the short two years since its establishment, the FCC-ONAP has assisted Tribal Nations by ensuring that their voices are included and present in all relevant FCC proceedings and initiatives; and

WHEREAS, the FCC-ONAP does not have an annual dedicated budget and this in turn results in a heightened level of uncertainty as to whether the FCC's engagement with Tribal Nations will continue on critical, current and pressing communications matters; and

WHEREAS, NCAI previously adopted Resolutions RAP-10-006 and MKE-11-006 in support of dedicated funding for the FCC-ONAP, recognizing the integral role the FCC-ONAP plays in ensuring tribes are included in FCC proceedings and initiatives.

NCAI HEADQUARTERS
1516 P Street, N.W.
Washington, DC 20005
202.466.7767
202.466.7797 fax
www.ncai.org

NOW THEREFORE BE IT RESOLVED, that the leadership of Tribal Nations through the NCAI, implore the FCC to prioritize and fully fund the FCC-ONAP on a dedicated and annual basis so that it will continue the progress of its important work in engaging Tribal Nations on current and critical communications issues; and

BE IT FURTHER RESOLVED, that NCAI reaffirms Resolutions RAP-10-006 and MKE-11-006 in support of funding and fully staffing the FCC-ONAP; and

BE IT FURTHER RESOLVED, NCAI urges the FCC to dedicate $2 million annually to the FCC-ONAP budget in support of the FCC's outreach, training, and consultation with Tribal Nations; and

BE IT FINALLY RESOLVED, that this resolution shall be the policy of NCAI until it is withdrawn or modified by subsequent resolution.

CERTIFICATION

The foregoing resolution was adopted by the General Assembly at the 2012 Annual Session of the National Congress of American Indians, held at the Sacramento Convention Center from October 21-26, 2012 in Sacramento, California, with a quorum present.

Jefferson Keel
President

ATTEST:

Edward K. Thomas
Recording Secretary

NATIONAL CONGRESS OF AMERICAN INDIANS

The National Congress of American Indians
Resolution #SAC-12-034

TITLE: Promoting Tribal Nation Access and Use of Spectrum for Communications Services

WHEREAS, we, the members of the National Congress of American Indians of the United States, invoking the divine blessing of the Creator upon our efforts and purposes, in order to preserve for ourselves and our descendants the inherent sovereign rights of our Indian nations, rights secured under Indian treaties and agreements with the United States, and all other rights and benefits to which we are entitled under the laws and Constitution of the United States, to enlighten the public toward a better understanding of the Indian people, to preserve Indian cultural values, and otherwise promote the health, safety and welfare of the Indian people, do hereby establish and submit the following resolution; and

WHEREAS, the National Congress of American Indians (NCAI) was established in 1944 and is the oldest and largest national organization of American Indian and Alaska Native tribal governments; and

WHEREAS, approximately 90% of Native Americans living in Indian Country do not have access to Internet connectivity and the economic, cultural and human significance of that fact cannot be underestimated; and

WHEREAS, connecting Indian Country to broadband with the rest of the world can reverse centuries of neglect and isolation and enable Tribal Nations to shape the future health and welfare of their communities with critical communications infrastructure; and

WHEREAS, broadband has the potential to assist Native American people in securing their rightful place in a world economy of ideas and opportunities; and

WHEREAS, access to currently licensed spectrum is absolutely necessary as many communities and entire Tribal Nations have not seen the full benefit of the services that could and should be provided on these licenses; and

WHEREAS, the Federal Communications Commission (FCC), Office of Native Affairs and Policy (FCC-ONAP) is well positioned and respected throughout Indian Country to consult and coordinate with Tribal Nations as the FCC acts on important proceedings that impact Tribal Nations; and

WHEREAS, with the creation of the Tribal Mobility Fund together with the Connect America Fund provisions requiring engagement with Tribal Nations on many important broadband deployment issues, Tribal Nations are still waiting for the promulgation of important rules to increase tribal access to spectrum; and

NCAI HEADQUARTERS
1516 P Street, N.W.
Washington, DC 20005
202.466.7767
202.466.7797 fax
www.ncai.org

WHEREAS, the attached letter dated July 19, 2012, NCAI President Jefferson Keel filed a letter to the FCC under WT Docket No. 11-40, highlighting critical issues for the FCC to act to increase tribal access to commercial wireless spectrum.

NOW THEREFORE BE IT RESOLVED, that in order to fully understand the true availability of communications service on tribal lands, the FCC must, as a threshold matter, consult with Tribal governments to ensure Tribes are receiving good faith, responsive, fair treatment and deployment of broadband infrastructure in accordance with tribal sovereignty; and

BE IT FURTHER RESOLVED, that NCAI reaffirms Resolutions MKE-11-007 and LNK-12-007, and calls for the FCC to implement a Tribal Priority in the rules inclusive of commercial mobile radio services and wireless spectrum that can be used to deploy critical important and robust broadband services; and

BE IT FURTHER RESOLVED, that NCAI, by our membership, adopts the comments made in the attached letter from NCAI President Jefferson Keel to the FCC; and

BE IT FINALLY RESOLVED, that this resolution shall be the policy of NCAI until it is withdrawn or modified by subsequent resolution.

CERTIFICATION

The foregoing resolution was adopted by the General Assembly at the 2012 Annual Session of the National Congress of American Indians, held at the Sacramento Convention Center from October 21-26, 2012 in Sacramento, California, with a quorum present.

Jefferson Keel
President

ATTEST:

Edward K. Thomas
Recording Secretary

SAC-12-034 Attachment

NATIONAL CONGRESS OF AMERICAN INDIANS

July 19, 2012

The Honorable Julius Genachowski
Chairman
Federal Communications Commission
445 12th Street SW
Washington, DC 20554

RE: **Emphasizing the Importance of a Tribal Priority to Spectrum Licenses (*Improving Communication Services for Native Nations by Promoting Greater Utilization of Spectrum Over Tribal Lands*, WT Docket No. 11-40)**

Dear Chairman Genachowski,

On behalf of the National Congress of American Indians (NCAI), I would like to extend to you my gratitude for speaking with NCAI's Board during our 2012 Executive Council Winter Session in early March. At that time I expressed how important and critical the need throughout Indian Country is for fully licensed wireless spectrum. As your Commission has acknowledged, wireless services are sorely lacking in many parts of Indian Country. In many places, wireless services are not a matter of convenience, but a matter of necessity, and often life or death.

NCAI appreciated the Commission launching the Spectrum for Tribal Lands Rulemaking. Tribal Nations need access to spectrum that was licensed long ago to companies that have failed to build out to communities on Tribal Lands. Much needed rule changes will serve tribal needs in these least connected regions of the country.

It is not only a matter of need but also a matter of efficiency in the use of this important resource, and especially in those many instances where the spectrum is not being used for the benefit of our communities. Having a wireless license is not the same as ownership of property, and such licenses should be utilized by those who are willing and able to deploy wireless services for the public good.

FCC Docket No. 11-40 – In the Matter of Improving Communications and Utilization of Spectrum Over Tribal Lands

NCAI and Native Public Media (NPM) submitted joint reply comments under WT Docket No. 11-40, "Improving Communication Services for Native Nations by Promoting Greater Utilization of Spectrum Over Tribal Lands" on June 20, 2011.[i] In the joint reply comments, NCAI and NPM highlighted a variety of measures the FCC could take to promote the deployment of wireless services over Tribal Lands, which included:

1) Expanding the Tribal Priority to advanced wireless services;
2) Establishing a build or divest process for spectrum use over Tribal Lands;
3) Establishing a formal and mandatory negotiation process between both current and future licensees and Tribal governments;

SAC-12-034 Attachment

4) The inclusion of a demonstration of service requirement as part of any Tribal-lands safe harbor;
5) Significant modification of the Tribal Lands Bidding Credit; and
6) The adoption of well defined eligibility criteria for Tribes to obtain spectrum, as set forth in the Notice of Proposed Rulemaking.[ii]

I respectfully submit this letter to reemphasize certain requirements for success that were previously addressed in NCAI's and NPM's joint reply comments to Docket No. 11-40. It has been well over a year since the FCC released its Notice of Proposed Rulemaking on this issue, and many tribes across the country are still waiting for the promulgation of these important rules to increase tribal access to spectrum.

Reemphasizing the Need to Increase Tribal Access to Spectrum
With the creation of the Tribal Mobility Fund together with the Connect America Fund provisions requiring engagement with Tribal Nations on many important broadband deployment issues, many pieces are in place to address the Digital Divide in Indian Country. Now is the time to put the most important piece in place – Spectrum priority.

With this in mind, I urgently encourage you to take the next steps and promulgate rules that the FCC has proposed to create new opportunities for Tribal Nations to access spectrum. Specifically, I urge you to take immediate action to:

1. **Implement a Tribal Priority in the rules inclusive of commercial mobile radio services, and wireless spectrum that can be utilized to deploy critical important and robust broadband services.** We look forward to future licensing actions or auction opportunities to license areas that support tribal lands and governmental priorities.

2. **Create rules that strengthen the structure of negotiations with existing licensed companies, and strengthen the ability of Tribal Nations to initiate and participate in these negotiations.** Access to currently licensed spectrum is absolutely necessary as many communities, and indeed entire Tribal Nations, have never seen the full benefit of the services that could and should be provided on these licenses. Adoption and utilization of broadband services cannot occur until these services are available on tribal lands.

3. **Ensure that there is good faith, responsiveness, and continuity in negotiations.** As part of the fiduciary trust responsibility that exists between the federal government and Tribal Nations, it is critical that the FCC act in accordance with the best interest of tribes. While NCAI supports FCC initiatives to ensure that industry entities must 'meaningfully engage' with tribal governments, the FCC should remain involved in these negotiations to ensure tribes are receiving fair treatment and deployment of broadband infrastructure is occurring in accordance with tribal sovereignty, FCC expectations, and community needs.

4. **Minimize the ability of parties to present ultimatums in negotiations, and one-sided demands.** In accordance with Executive Order 13175, "Consultation and Coordination with Tribal Governments",[iii] and President Obama's recent Executive Order, "Accelerating Broadband Infrastructure Deployment",[iv] the FCC should develop and implement mechanisms for tribal recourse when and if negotiations with industry entities deteriorate.

SAC-12-034 Attachment

5. **Create opportunities for these negotiations to recognize the unique geography, economies, and multiple-use needs of many Tribal Nations.** Coordinated efforts between tribes, their respective governments, and industry entities are essential to bridging the Digital Divide in Indian Country. Planning and feasibility studies for infrastructure deployment should take into consideration the needs of the community. The recent Connect America Fund 'Tribal Engagement Provisions' provide a foundation for this to occur by requiring eligible telecommunications carriers (ETCs) to meaningfully engage with tribal governments in their service areas. NCAI supports these provisions and encourages the FCC to enforce these provisions on ETCs serving tribal lands.

6. **Further investigate Best Practices based on stronger rules and experiences surrounding negotiations and tribal engagement under the Connect America Fund.** The new build out measures for Tribal Lands should be evaluated in coordination with Tribal Nations, and involve the carriers who are only now beginning to address tribes within their service areas. These measures should be tailored specifically to the needs of different regions and address the unique aspects of tribal governments and lands, while taking into consideration the needs of each individual Tribal Nation.

7. **Implement a "Build or Divest" program in the context of close coordination and consultation between Tribal Nations and the FCC.** We recognize this is a controversial point, but until the Commission takes action to enforce actual build out onto tribal lands, we will not solve the specific issues in each corner of Indian Country. Recipients of CAF funds should be legally obligated to address these specific issues and needs. Additionally, the concept of "Build or Divest" is similar to the incentive auctions authorized by Congress in the Middle Class Tax Relief and Job Creation Act of 2012 as a way to encourage the voluntary relinquishment of currently licensed spectrum.

Implementation of Tribal Priority to Spectrum Would Benefit Tribal Nations Today

In closing, I reiterate our offer to assist in any further outreach efforts to tribes for the purposes of consultation and coordination with Tribal Nations as the Commission acts in this important proceeding. With the FCC Office of Native Affairs and Policy well-positioned and well-respected throughout Indian Country, the Commission's presence across Indian Country is deeply appreciated and much needed. With the many new opportunities for Tribal Nations that the Commission has created, it is time to implement a tribal spectrum priority and bring meaningful robust wireless services to tribal lands.

Therefore, the Commission should act now to increase access to spectrum over tribal lands. A Tribal Priority to spectrum would enable American Indian tribes and Alaska Native villages to develop their own robust wireless services if they so choose or to empower tribes to develop their own regulatory authorities to engage with industry officials in the planning and deployment of robust wireless services.

Sincerely,

Jefferson Keel
President, National Congress of American Indians

SAC-12-034 Attachment

Enclosed: Resolution #LNK-12-007 – Spectrum Allocation and the Low Power FM Radio (LPFM) 'Tribal Priority'

Cc. Geoffrey Blackwell
Chief, Office of Native Affairs and Policy
Federal Communications Commission

[i] *See* Joint Reply Comments of Native Public Media and the National Congress of American Indians. WT Docket No. 11-40. "In the Matter of Improving Communication Services for Native Nations by Promoting Greater Utilization of Spectrum Over Tribal Lands." Submitted to the FCC Electronic Comment Filing System on June 20, 2011. Posted on June 21, 2011. *Available* at http://apps.fcc.gov/ecfs/document/view?id=7021688922.
[ii] *Id.*
[iii] *See* Executive Order No. 13175, "Consultation and Coordination with Tribal Governments. 65 Federal Register 218. November 9, 2000. *Available* at http://www.gpo.gov/fdsys/pkg/FR-2000-11-09/pdf/00-29003.pdf.
[iv] *See* President Obama Executive Order, "Accelerating Broadband Infrastructure Deployment". June 14, 2012. Available at http://www.whitehouse.gov/the-press-office/2012/06/14/executive-order-accelerating-broadband-infrastructure-deployment.

AFFILIATED TRIBES OF NORTHWEST INDIANS
ATNI

2013 Winter Convention
Grand Mound, WA

RESOLUTION #13 - 09

"EFFORT TO GATHER INFORMATION ON FORMING TRIBAL TELECOMMUNICATIONS COMPANIES AND EDUCATE MEMBER TRIBES"

PREAMBLE

We the members of the Affiliated Tribes of Northwest Indians of the United States, invoking the divine blessing of the Creator upon our efforts and purposes, in order to preserve for ourselves and our descendants rights secured under Indian Treaties, Executive Orders, and benefits to which we are entitled under the laws and constitution of the United States and several states, to enlighten the public toward a better understanding of the Indian people, to preserve Indian cultural values, and otherwise to promote the welfare of the Indian people, do hereby establish and submit the following resolution:

WHEREAS, the Affiliated Tribes of Northwest Indians (ATNI) are representatives of and advocates for national, regional, and specific tribal concerns; and

WHEREAS, ATNI is a regional organization comprised of American Indians/Alaska Natives and tribes in the states of Washington, Idaho, Oregon, Montana, Nevada, Northern California, and Alaska; and

WHEREAS, the health, safety, welfare, education, economic and employment opportunity, and preservation of cultural and natural resources are primary goals and objectives of the ATNI; and

WHEREAS, in 2012 the Federal Communications Commission ("FCC") issued major reforms to an $8 Billion annual subsidy fund known as the Universal Service Fund ("USF"); and

AFFILIATED TRIBES OF NORTHWEST INDIANS **RESOLUTION #13 - 09**

WHEREAS, the FCC Office of Native Affairs and Policy is actively encouraging tribes to seriously consider becoming Eligible Telecommunications Carriers ("ETC") by leveraging new, favorable policy within the USF which includes tens of millions of dollars annually in tribal set-asides, tribal preference, bidding credits, capital funding and operational subsidies; and

WHEREAS, the reformed USF and the new subfund known as the Connect America Fund ("CAF") are designed to support phone and broadband infrastructure and service in rural and high-cost markets; and

WHEREAS, communications services on tribal lands and areas are commonly subpar, with incumbent communications carriers (service providers) typically refusing or unresponsive in addressing issues of coverage, quality and cost; and

WHEREAS, the reform of the USF provides tribes with a substantial opportunity to take control of communications with substantial impact on the economic and social welfare; and

WHEREAS, certain opportunities within USF and CAF may be time sensitive and it is prudent for ATNI to assemble information about the USF and CAF and educate its member tribes about these opportunities quickly; now

THEREFORE BE IT RESOLVED, that ATNI shall take on an effort to assemble information and resources regarding the USF/CAF and formulate a prudent method by which to educate, support and encourage its member tribes to consider the opportunity provided by the USF reform.

CERTIFICATION

The foregoing resolution was adopted at the 2013 Winter Convention of the Affiliated Tribes of Northwest Indians, held at Great Wolf Lodge, Grand Mound, Washington, on January 28 – 31, 2013 with a quorum present.

Fawn Sharp, President

Norma Jean Louie, Secretary

2013 Winter Convention
Grand Mound, WA

RESOLUTION #13 - 10

"INTENT TO SUBMIT COMMENTS TO FCC ON ISSUES OF WIRELESS SPECTRUM BY ATNI AND ENCOURAGE INDIVIDUAL TRIBAL COMMENT SUBMISSION"

PREAMBLE

We the members of the Affiliated Tribes of Northwest Indians of the United States, invoking the divine blessing of the Creator upon our efforts and purposes, in order to preserve for ourselves and our descendants rights secured under Indian Treaties, Executive Orders, and benefits to which we are entitled under the laws and constitution of the United States and several states, to enlighten the public toward a better understanding of the Indian people, to preserve Indian cultural values, and otherwise to promote the welfare of the Indian people, do hereby establish and submit the following resolution:

WHEREAS, the Affiliated Tribes of Northwest Indians (ATNI) are representatives of and advocates for national, regional, and specific tribal concerns; and

WHEREAS, ATNI is a regional organization comprised of American Indians/Alaska Natives and tribes in the states of Washington, Idaho, Oregon, Montana, Nevada, Northern California, and Alaska; and

WHEREAS, the health, safety, welfare, education, economic and employment opportunity, and preservation of cultural and natural resources are primary goals and objectives of the ATNI; and

WHEREAS, the Federal Communications Commission ("FCC") is currently working to reform existing policy and issue new policy on spectrum and requires comments on the record (for rulemaking) representing the interests of tribes in order to build favorable policy; and

WHEREAS, the FCC Office of Native Affairs and Policy has actively encouraged tribal councils to submit comments, stories, and proposed policy on wireless spectrum issues including; sovereignty and ownership of spectrum over tribal lands, impacting currently allocated/owned spectrum, the lease of spectrum for effective use and allocation of future spectrum; and

WHEREAS, wireless spectrum is key to 21st Century communications (phone, internet, mobility) and are critical to the operation and growth of education, public safety, government services and economic development; and

WHEREAS, communications services on tribal lands and areas are commonly subpar, with incumbent communications carriers (service providers) typically refusing or unresponsive in addressing issues of coverage, quality and cost; and

WHEREAS, wireless spectrum is a natural resource (like timber, fish and minerals) and should be protected by ATNI and member tribes in a similar fashion as other natural resources, acknowledging its role as critical infrastructure with substantial impact on the economic and social welfare of tribes; and

WHEREAS, immediate and regular involvement by ATNI and its member tribes are critical to the foundation of favorable policy by the FCC and maintaining tribal sovereignty; now

THEREFORE BE IT RESOLVED, that ATNI shall submit formal recommendation(s) on issues of spectrum policy with signatures of ATNI member tribes for strong consideration, and, as appropriate, introduction and enactment, in the Federal Communications Commission's rulemaking(s), for the policy reasons cited above and to further tribal sovereignty; and

BE IT FURTHER RESOLVED, that ATNI shall also promote and support efforts by its member tribes to submit individual comments to further strengthen the record of comment by which FCC creates policy.

CERTIFICATION

The foregoing resolution was adopted at the 2013 Winter Convention of the Affiliated Tribes of Northwest Indians, held at Great Wolf Lodge, Grand Mound, Washington, on January 28 – 31, 2013 with a quorum present.

Fawn Sharp, President

Norma Jean Louie, Secretary

End Notes

[1] The work of the Office of Native Affairs and Policy encompasses all 566 federally recognized Tribes, which includes Alaska Native Villages. The work of the Office also encompasses the Hawaiian Home Lands – lands held in trust for native Hawaiians by the state of Hawaii, pursuant to the Hawaiian Homes Commission Act, 1920, Act July 9, 1921, 42 Stat. 108, *et seq.*, as amended.

[2] *See Connect America Fund; A National Broadband Plan for Our Future; Establishing Just and Reasonable Rates for Local Exchange Carriers; High-Cost Universal Service Support; Developing a Unified Intercarrier Compensation Regime; Federal-State Joint Board on Universal Service; Lifeline and Link-Up; Universal Service Reform—Mobility Fund;* WC Docket Nos. 10-90, 07-135, 05-337, 03-109, CC Docket Nos. 01-92, 96-45, GN Docket No. 09-51, WT Docket No. 10-208, Report and Order and Further Notice of Proposed Rulemaking, 26 FCC Rcd 17663 (2011) (*USF/ICC Transformation Order*); *pets. for review pending sub nom. In re: FCC 11-161*, No. 11- 9900 (10th Cir. filed Dec. 8, 2011); *see infra* A Renewed Regulatory Agenda – Universal Service Reform, pp. 20- 21.

[3] *See USF/ICC Transformation Order*, 26 FCC Rcd at 17868, paras. 636-37.

[4] The NEPA team is comprised of managers and staff members representing the Wireless Telecommunications Bureau, the Media Bureau, the Public Safety and Homeland Security Bureau, the Office of General Counsel, the Office of Native Affairs and Policy, and the Commission's Federal Preservation Officer. The focus of the NEPA team, which meets on a weekly basis, is on the effects of construction of communications facilities by or for the use of the Commission's licensees under federal environmental statutes, including the National Environmental Policy Act (NEPA), the Endangered Species Act (ESA), and the National Historic Preservation Act (NHPA).

[5] *Nationwide Programmatic Agreement for the Collocation of Wireless Antennas*, 47 C.F.R. Pt.1, App.B (2001); *Nationwide Programmatic Agreement Regarding the Section 106 National Historic Preservation Act Review Process*, WT Docket No. 03-128, Report and Order, 20 FCC Rcd 1073 (2004); Voluntary Best Practices for Expediting the Process of Communications Tower and Antenna Siting Review pursuant to Section 106 of the National Historic Preservation Act (adopted Oct. 25, 2004).

[6] For example, the Commission's Tower Construction Notification System (TCNS) is a private, highly developed, and password protected online notification system that facilitates the interaction between tower constructors and Tribal representatives on proposed tower constructions.

[7] *Policies to Promote Rural Radio Service and to Streamline Allotment and Assignment Procedures*, MB Docket No. 09-52, Third Report and Order, 26 FCC Rcd 17642 (2011) (*Rural Radio Third Report and Order*); *Policies to Promote Rural Radio Service and to Streamline Allotment and Assignment Procedures*, MB Docket No. 09-52, First Report and Order and Further Notice of Proposed Rule Making, 25 FCC Rcd 1583, 1586 at para. 5 (2010) (*Rural Radio First Report and Order*).

[8] *Extending Wireless Telecommunications Services to Tribal Lands*, WT Docket No. 99-266, Report and Order and Further Notice of Proposed Rule Making, 15 FCC Rcd 11794, 11798 (2000); *see also Connecting America: The National Broadband Plan*, prepared by the staff of the Federal Communications Commission, March 2010 (National Broadband Plan) at 152, Box 8-4; *see also USF/ICC Transformation Order*, 26 FCC Rcd at 17868, para. 636.

[9] *Improving Communications Services for Native Nations*, CG Docket No. 11-41, Notice of Inquiry, 26 FCC Rcd 2672, 2673 at para. 1 (2011) (*Native Nations NOI*).

[10] TRACI L. MORRIS, NATIVE PUBLIC MEDIA, & SASCHA D. MEINRATH, NEW AMERICA FOUNDATION, NEW MEDIA, TECHNOLOGY AND INDIAN USE IN INDIAN COUNTRY (2009) (*NPM/NAF New Media Study*).

[11] From testimony available at http://transition.fcc.gov/Panel_Discussions/Teleservice_ reservations/tr-newmx.txt (Pages 41-42).

[12] From testimony available at http://transition.fcc.gov/Panel_Discussions/Teleservice_ reservations/march23/32399fcc.txt and http://transition.fcc.gov/Panel_Discussions /Teleservice_reservations/tr-newmx.txt

[13] *See Establishing a Government-to-Government Relationship with Indian Tribes*, Policy Statement, 16 FCC Rcd 4078, 4080-81 (2000) (*Tribal Policy Statement*).

[14] *See Federal-State Joint Board on Universal Service; Promoting Deployment and Subscribership in Unserved and Underserved Areas, Including Tribal and Insular Areas*, CC Docket No. 96-45, Twelfth Report and Order, Memorandum Opinion and Order and Further Notice of Proposed Rulemaking, 15 FCC Rcd 12208 (2000) (*Twelfth Report and Order*).

[15] *Twelfth Report and Order*, 15 FCC Rcd at 12261, para. 108.

[16] *See Extending Wireless Telecommunications Services to Tribal Lands*, WT Docket No. 99-266, Report and Order and Further Notice of Proposed Rulemaking, 15 FCC Rcd 11794 (2000).

[17] *See* 16 U.S.C. § 470f. Section 106 generally requires federal agencies to consider the effects of their proposed undertakings on historic properties. Section 101 (d)(6) specifically governs how historic properties of traditional cultural and religious importance to federally recognized Tribes are to be considered in the Section 106 process. 16 U.S.C. § 470a(d)(6).

[18] *Nationwide Programmatic Agreement for the Collocation of Wireless Antennas*, 47 C.F.R. Pt.1, App.B.

[19] *Nationwide Programmatic Agreement Regarding the Section 106 National Historic Preservation Act Review Process*, WT Docket No. 03-128, Report and Order, 20 FCC Rcd 1073 (2004); *see also Nationwide Programmatic Agreement Regarding the Section 106 National Historic Preservation Act Review Process*, 47 C.F.R. Pt.1, App.C.

[20] Federal Communications Commission, Wireless Telecommunications Bureau, Towers and Antennas: TCNS/E106, http://wireless.fcc.gov/outreach/index.htm?job=tower_notification (last visited Mar. 19, 2013).

[21] Voluntary Best Practices for Expediting the Process of Communications Tower and Antenna Siting Review pursuant to Section 106 of the National Historic Preservation Act (adopted Oct. 25, 2004).

[22] Memorandum of Understanding between the Federal Communications Commission and the United South and Eastern Tribes, Inc., Regarding Recommended Best Practices and the Section 106 Process (adopted Feb. 3, 2004).

[23] *High-Cost Universal Service Support; Federal-State Joint Board on Universal Service*; WC Docket No. 05-337, CC Docket No. 96-45, Order, 23 FCC Rcd 8834 (2008) (*Interim Cap Order*).

[24] Chairman Kennard's speech is available at http://transition.fcc.gov/Speeches/Kennard /2000/spwek021.doc.

[25] Press Release, Federal Communications Commission, FCC Announces Indian Tele communications Initiative (Apr. 30, 2002), *available at* http://transition.fcc.gov /Bureaus/CGB/News_Releases/2002/nrcg0202.html (last visited Mar. 19, 2013).

[26] *See generally National Broadband Plan.* Tribal-specific recommendations included the creation of a Tribal Broadband Fund to support sustainable broadband deployment and adoption on Tribal lands (Recommendation 8.18); the creation of an FCC-Tribal Broadband

Task Force (Recommendation 9.14); and the establishment of Tribal seats on the Universal Service Joint Board and the Universal Service Administrative Company's Board of Directors (Recommendation 9.14).

[27] Commenters included, for example, the National Congress of American Indians, the California Association of Tribal Governments, and Native Public Media.

[28] *National Broadband Plan* at 184.

[29] *Establishment of the Office of Native Affairs and Policy in the Consumer and Governmental Affairs Bureau*, Order, 25 FCC Rcd 11104 (2010) (*ONAP Order*).

[30] *Policies to Promote Rural Radio Service and to Streamline Allotment and Assignment Procedures*, MB Docket No. 09-52, Second Report and Order, First Order on Reconsideration, and Second Further Notice of Proposed Rule Making, 26 FCC Rcd 2556 (2011) (*Rural Radio Second Report and Order*); Improving Communications Services for Native Nations by Promoting Greater Utilization of Spectrum over Tribal Lands, WT Docket No. 11-40, Notice of Proposed Rulemaking, 26 FCC Rcd 2623 (2011) (*Spectrum over Tribal Lands NPRM*); *Native Nations NOI*, 26 FCC Rcd 2672.

[31] *See infra* A Renewed Regulatory, pp. 18-24.

[32] *See, e.g., Seminole Nation v. United States*, 316 U.S. 286, 296 (1942) (citing *Cherokee Nation v. State of Georgia*, 30 U.S. 1 (1831); *United States v. Kagama*, 118 U.S. 375 (1886); *Choctaw Nation v. United States*, 119 U.S. 1 (1886); *United States v. Pelican*, 232 U.S. 442 (1914); *United States v. Creek Nation*, 295 U.S. 103 (1935); *Tulee v. State of Washington*, 315 U.S. 681 (1942).

[33] *See, e.g., United States v. Mitchell*, 463 U.S. 206 (1983).

[34] *See, e.g., The Indian Financing Act of 1974*, 25 U.S.C. § 1451 (1974); *The Indian Self-Determination and Education Assistance Act of 1975*, 25 U.S.C. § 450 (1975); *The Indian Civil Rights Act of 1968*, 25 U.S.C. § 1301 (1968); *see also White Mountain Apache Tribe v. Bracker*, 448 U.S. 136, 142 (1980); *New Mexico v. Mescalero Apache Tribe*, 462 U.S. 324, 334 (1983).

[35] *Tribal Policy Statement*, 16 FCC Rcd at 4080-81.

[36] *Id.*

[37] *Native Nations NOI*, 26 FCC Rcd at 2676, para. 5.

[38] *ONAP Order*, 25 FCC Rcd at 11104, para. 1.

[39] *Native Nations NOI*, 26 FCC Rcd at 2679, para. 12.

[40] *See generally Native Nations NOI*, 26 FCC Rcd 2672.

[41] *See generally USF/ICC Transformation Order*, 26 FCC Rcd 17663 (for example, the Commission established Tribal government engagement obligations and created the Tribal Mobility Fund).

[42] *USF/ICC Transformation Order*, 26 FCC Rcd at 17819, para. 481.

[43] *USF/ICC Transformation Order*, 26 FCC Rcd at 17858, para. 604.

[44] *Id.* at 17868, para. 637; *Office of Native Affairs and Policy, Wireless Telecommunications Bureau, and Wireline Competition Bureau Issue Further Guidance on Tribal Government Engagement Obligation Provisions of the Connect America Fund*, WC Docket No. 10-90 et al., Public Notice, 27 FCC Rcd 8176, 8178 (Wireline Comp. Bur., Wireless Telecomm. Bur., Off. of Native Aff. & Pol'y 2012) (*Further Guidance Public Notice*).

[45] *USF/ICC Transformation Order*, 26 FCC Rcd at 17741-47, paras. 210-26.

[46] *Connect America Fund High-Cost Universal Support*, WC Docket Nos. 10-90, 05-37, Order, 27 FCC Rcd 4235 at Appendix A, para. 101 (Wireline Comp. Bur. 2012) (*Benchmarks Order*).

[47] *Lifeline and Link Up Reform and Modernization; Lifeline and Link Up; Federal-State Joint Board on Universal Service; Advancing Broadband Availability through Digital Literacy Training;* WC Docket Nos. 11-42, 03-109, 12 23, CC Docket No. 96-45, Report and Order and Further Notice of Proposed Rulemaking 27 FCC Rcd 6656, 6761 2, at para. 245 (2012) (*Lifeline Modernization Order*) (maintaining, at the present time, enhanced Link Up support for those ETCs that also receive high-cost support on Tribal lands).

[48] *Lifeline and Link Up Reform and Modernization*, WC Docket No. 11-42, Order, 27 FCC Rcd 15842 (2012) (*Lifeline Pilot Program Order*).

[49] *See generally Spectrum over Tribal Lands NPRM*, 26 FCC Rcd 2623.

[50] *Spectrum over Tribal Lands NPRM,* 26 FCC Rcd at 2628, para. 10.

[51] *See generally Rural Radio First Report and Order*, 25 FCC Rcd 1583.

[52] *See generally Rural Radio Second Report and Order*, 26 FCC Rcd at 2556.

[53] *See generally Rural Radio Third Report and Order,* 26 FCC Rcd at 17642.

[54] *See Creation of a Low Power Radio Service; Amendment of Service and Eligibility Rules for FM Broadcast Translator Stations*, MM Docket No. 99-25, MB Docket No. 07-172, RM 11338, Fifth Order on Reconsideration and Sixth Report and Order, 27 FCC Rcd 15402 (2012) (*LPFM Order*).

[55] *See infra* Acknowledgement of the FCC-Native Nations Broadband Task Force, pp. 62-63.

[56] *Application for the FCC's Broadband Adoption Lifeline Pilot Program*, WC Docket No. 11-42 at 2 (filed Jul. 9, 2012) (*HTI Broadband Pilot Application*).

[57] BUREAU OF INDIAN AFFAIRS, THE AMERICAN INDIAN POPULATION AND LABOR FORCE REPORT (2005), available at http://www.bia.gov/cs/groups/public/documents/text/idc-001719.pdf (last visited Mar. 19, 2013).

[58] The Hopi Tribe, Tribal Government, http://www.hopi-nsn.gov/TribalGovernment/tabid/64/Default. aspx (last visited Mar. 19, 2013).

[59] Hopi Telecommunications, Inc., History, http://www.hopitelecom.com/about-history.php (last visited Mar. 19, 2013).

[60] *Id.*

[61] *HTI Broadband Pilot Application* at 2.

[62] *HTI Broadband Pilot Application* at 2.

[63] *Id.*

[64] Hopi Telecommunications, Inc., Ex Parte Presentation, WC Docket Nos. 10-90, 07-135, 05-337, 03-109, GN Docket No. 09-51, CC Docket Nos. 01-92, 96-45, Oct. 14, 2011 (*HTI 2011 Ex Parte*); *HTI Broadband Pilot Application* at 8.

[65] Hopi Telecommunications, Inc., Comments, WT Docket 10-208, WC Docket Nos. 10-90, 07-135, 05-337, 03- 109, GN Docket No. 09-51, CC Docket Nos. 01-92, 96-45, Jan. 18, 2012 at 3 (*HTI 2012 Comments*).

[66] *HTI Broadband Pilot Application* at 7.

[67] *See generally Tribal Policy Statement*, 16 FCC Rcd 4078.

[68] *HTI 2011 Ex Parte* at 3; *HTI 2012 Comments* at 7.

[69] *Benchmarks Order,* 27 FCC Rcd at Appendix A, para. 101.

[70] *See generally Further Guidance Public Notice*, 27 FCC Rcd 8176.

[71] *ONAP Order*, 25 FCC Rcd at 11104, para. 1.

[72] The 17 Tribes connected via the Tribal Digital Village system are as follows: Barona Band of Mission Indians, Cahuilla Band of Mission Indians, Campo Kumeyaay Nation, Ewiiaapaayp Band of Kumeyaay Indians, Iipay Nation of Santa Ysabel, Jamul Indian Village, La Jolla Band of Luiseño Indians, La Posta Band of Mission Indians, Los Coyotes Band of Cahuilla and Cupeño Indians, Manzanita Band of Diegueño Mission Indians, Mesa

Grande Band of Mission Indians, Pala Band of Mission Indians, Pauma Band of Luiseño Indians, Rincon Band of Luiseño Indians, San Pasqual Band of Mission Indians, Soboba Band of Luiseño Indians, and Tachi Yokut Tribe.

[73] American Recovery and Reinvestment Act of 2009, Pub. L. No. 111-5, 123 Stat. 115, 516, § 6001(k)(2)(D) (Recovery Act).

[74] *Unlicensed Operation in the TV Broadcast Bands; Additional Spectrum for Unlicensed Devices Below 900 MHz and in the 3 GHz Band;* ET Docket Nos. 04-186, 02-380, Second Memorandum Opinion and Order, 25 FCC Rcd 18661 (2010) (*White Spaces Order*).

[75] Letter from Jefferson Keel, President, National Congress of American Indians, to Julius Genachowski, Chairman, Federal Communications Commission, WT Docket No. 11-40, July 19, 2012 (*President Keel Ex Parte Letter*).

[76] The Confederated Tribes of the Colville Reservation, Demographics, http://www.colvilletribes.com/demographics.php (last visited Mar. 19, 2013).

[77] *See USF/ICC Transformation Order*, 26 FCC Rcd at 17868, para. 637, n.1054; *see also Further Guidance Public Notice*, 27 FCC Rcd at 8178-79, para. 8.

[78] *See supra* The Commission's Tribal Agenda before the Office of Native Affairs and Policy, pp. 8-13; *see also Tribal Policy Statement*, 16 FCC Rcd at 4078-82.

[79] 47 U.S.C. § 307(b).

[80] *Rural Radio First Report and Order*, 25 FCC Rcd at 1584, 1587-8, paras. 1, 8.

[81] *Rural Radio Third Report and Order,* 26 FCC Rcd at 17642-4, paras. 1-3.

[82] *Id.* at 17642-3, para. 1.

[83] Hualapai Tribe, About Hualapai, http://hualapai-nsn.gov/about-2 (last visited Mar. 19, 2013).

[84] *Id.*

[85] *Petition for Rulemaking to Amend Section 73.202(b), Table of Allotments, FM Broadcast Stations*, MB Docket No. 12-374, RM-11687 (filed Oct. 11, 2012) (*Hualapai Petition*).

[86] Julie Cart, *Tribe's Canyon Skywalk Opens One Deep Divide*, L.A. TIMES, Feb. 11, 2007, *available at* http://www.latimes.com/la-na-skywalk11feb11,0,6435928.story (last visited Mar. 19, 2013).

[87] *Hualapai Petition* at 3.

[88] *Id.*

[89] *Id.*

[90] *Peach Springs, Arizona*, MB Docket No. 12-374, RM-11687, Notice of Proposed Rulemaking, 27 FCC Rcd 15954 (Media Bur. 2012).

[91] *Peach Springs, Arizona*, MB Docket No. 12-374, RM-11687, Report and Order (Media Bur. rel. Mar. 1, 2013). Simultaneous to the release of this order, the Audio Division added FM Channel 297A at Crownpoint, New Mexico, as a Tribal Allotment in response to a petition for rulemaking filed by Navajo Technical College. *Crownpoint, New Mexico*, MB Docket 12-261, RM-11677, Report and Order (Media Bur. rel. Mar. 1, 2013).

[92] *See* EPCH Radio – Hualapai Tribal Radio, http://www.epchradio.com (last visited Mar. 19, 2013).

[93] *Rural Radio Third Report and Order*, 26 FCC Rcd at 17643-44, para. 3.

[94] *Rural Radio Third Report and Order*, 26 FCC Rcd at 17650-1, para. 16.

[95] Native Public Media, Native Radio Stations, http://secure.nativepublicmedia.org/radio (last visited Mar. 19, 2013).

[96] Native Public Media and the National Congress of American Indians, Reply to Comments, MB Docket No. 09-52, RM-11528, Aug. 11, 2009.

[97] Standing Rock Sioux Tribe, Community Environmental Profile, http://www.standingrock.org/communityProfile/ (last visited Mar. 19, 2013).

[98] *Telecommunications Carriers Eligible for Universal Service Support; Standing Rock Telecommunications, Inc. Petition for Designation as an Eligible Carrier; Petition of Standing Rock Telecommunications, Inc. to Redefine Rural Service Areas; Petition for Reconsideration of Standing Rock Telecommunications, Inc.'s Designation as an Eligible Telecommunications Carrier on the Standing Rock Sioux Reservation;* WC Docket No. 09-197, Memorandum Opinion and Order on Reconsideration, 26 FCC Rcd 9160, n. 36 (2011) (*Standing Rock Order Part II*).

[99] Standing Rock Sioux Tribe, Community Environmental Profile, http://www.standingrock.org /communityProfile/ (last visited Mar. 19, 2013).

[100] Standing Rock Sioux Tribe, Community Environmental Profile, http://www.standingrock.org /communityProfile/ (last visited Mar. 19, 2013).

[101] *Standing Rock Order Part II,* 26 FCC Rcd at 9160, para. 2; *Telecommunications Carriers Eligible for Universal Service Support; Standing Rock Telecommunications, Inc. Petition for Designation as an Eligible Telecommunications Carrier; Standing Rock Telecommunications, Inc.; Petition to Redefine Rural Service Areas;* WC Docket No. 09-197, Memorandum Opinion and Order, 25 FCC 12388 (Wireline Comp. Bur. 2010) (*Standing Rock Order Part I*).

[102] *Standing Rock Order Part II,* 26 FCC Rcd at 9165, para. 14.

[103] *Id.* at 9166-7, para. 19.

[104] *USF/ICC Transformation Order*, 26 FCC Rcd at 17836, paras. 530-31.

[105] *Id.* at 17797, para. 386.

[106] *Id.* at 17823, para. 491.

[107] *See President Keel Ex Parte Letter.*

[108] *Lifeline Modernization Order*, 27 FCC Rcd at 6767, para. 254. The Commission eliminated Link Up everywhere except on Tribal lands. Only ETCs receiving High-Cost support may also receive Link Up on Tribal lands.

In: Bridging the Digital Divide in Indian Country ISBN: 978-1-62948-170-8
Editors: R. Louton and J. Sullivan

Chapter 2

STATEMENT OF GEOFFREY BLACKWELL, CHIEF, OFFICE OF NATIVE AFFAIRS AND POLICY, FEDERAL COMMUNICATIONS COMMISSION. HEARING ON "INTERNET INFRASTRUCTURE IN NATIVE COMMUNITIES: EQUAL ACCESS TO E-COMMERCE, JOBS AND THE GLOBAL MARKETPLACE"*

Chairman Akaka, Vice Chairman Barrasso, and Members of the Committee, thank you for the opportunity to testify today about the importance of broadband infrastructures in Native Nations and Communities, and the Commission's efforts to work with Native leaders to find viable solutions.

The lack of all communications services in Indian Country is alarming. Our most recent reliable census data indicates that over 70 years of development and expansion of the telecommunications industry has resulted in only 67.9 percent of residents of Tribal lands enjoying basic telephone service. The statistics for broadband penetration are even more troubling -- less than 10 percent of residents of Native Nations have access to the lifeblood of our 21st century economy, educational opportunities, health care, and public safety.

---

* This is an edited, reformatted and augmented version of a statement, presented October 6, 2011 before the Senate Committee on Indian Affairs.

This past April, I told the Senate Commerce Committee what many on the Senate Committee on Indian Affairs already know -- that these too familiar statistics paint only part of the picture and behind them lurks a stark and complex reality. The negative impacts of history fell particularly hard on Tribal and Native Communities. One result of this history is an endemic lack of many critical infrastructures. In fact, almost no critical infrastructure has come to Tribal lands without federal investment, oversight, and regulation. Broadband opportunities can do much to level this history in bringing health care, education, jobs, and the opportunities of hope to Native Nations, but broadband must be available, accessible, and affordable to meet its great promise.

The purpose of the Office of Native Affairs and Policy is to change the approach to the communications problems of Native America. Our work with Native Nations is a new strategic partnership, one in which we effectuate and exercise the trust relationship that the Commission shares with Native Nations. There are numerous and comprehensive communications needs throughout Indian Country, and there is great diversity within those critical needs. The need for telemedicine is greatest for some Tribal Nations, while the needs for educational technology or public safety are paramount for others. In many Native places, Indian Reservations for instance, connectivity often occurs only in border towns and along major transportation routes crossing over Tribal lands. It is clear that one size fits none, and the enormity of our mission is vast. Changing our rules alone is not enough. Complex problems require new approaches and mechanisms, and active efforts both in Washington and in far into the field to develop and coordinate well thought-out solutions.

Created by a unanimous vote of the Commission on August 12, 2010 and implementing a recommendation of the National Broadband Plan, the Office of Native Affairs and Policy is now just over a year old. The Office is responsible for developing and driving a Tribal agenda at the Commission and serves as the Commission's primary point of contact on all Native issues. The Office is charged with bringing the benefits of a modern communications infrastructure to all Native communities by, among other things, ensuring robust government-to-government consultation with Federally-recognized Tribal governments and other Native organizations; working with Commissioners, Bureaus, and Offices, as well as with other government agencies and private organizations, to develop and implement policies for assisting Native communities; and ensuring that Native concerns and voices are considered in all relevant Commission proceedings and initiatives. Under Chairman Genachowski's leadership, and with the involvement of the entire

Commission and all of its Bureaus and Offices, there is a new way of doing Native business at the Commission, and Native Nations are central in that new paradigm.

The Efforts of the Office of Native Affairs and Policy

Our approach is to work together to identify and remove barriers to solutions and build models with Native Nations that engage their core community or anchor institutions. We seek to place Native Nations themselves in the center of those solutions, whether it is through actual self-provisioning of communications services or through new "Tribal-centric" or "Native-centric" methods of deployment with industry, public, or private partners. These models must respect the cultural values and sovereign priorities of Native communities and be infused with the local knowledge that will lead to better local involvement and opportunities for success. As Tribes govern with a unique understanding of their communities, their vested and active involvement is critically important to finding lasting solutions in their communities.

To fulfill its mission, the Office is fostering the Commission's ongoing government-togovernment dialogue with Native Nations by working directly with them to understand their needs and empower them to provide their own solutions. New opportunities must be created for Native Nations and those who work with them to find sustainable solutions. To fulfill our mission and transform the communications landscape, our Office cannot be just another outsider from Washington. Instead, the Office must be a knowledgeable and respected Indian Country *insider*. We must foster and maintain an expert understanding and familiarity with Tribal lands and Native Communities. Collectively, our four senior staff members have over 40 years of experience working in the trenches of the Commission and directly with Native Nations. We are adding to those ranks and we stand ready for the challenge.

One year ago, immediately upon being established, we hit the ground running. We actually rolled out the introduction of our new Office in Native America on a "listening tour," while at the same time working across the Commission to surface actions and proposals. We have continued with our commitment to working with Native leaders in their own reservations and homelands, where the problems actually exist. Side-by-side with our Native

Nation colleagues, we have "kicked the dirt" within numerous Native Nations, and discussed how we can help them with their development and deployment plans.

Several times, we have had to reset our phones and blackberries, log off and log back in, and set our out-of-office automatic reply messages to let folks know we are traveling in very unconnected regions.

Within our first year of operations, we traveled to and met with Tribal leaders in Arizona, California, Idaho, Montana, Nevada, New Mexico, North Dakota, Oklahoma, South Dakota, Utah, and Washington, as well as within Hawaiian Home Lands. Other remote and underserved areas of the country, including those within Alaska, are at the top of our future travel priorities. We logged thousands of miles and traveled to places where the Commission has never been before, experiencing the lack of connectivity from the other end of the digital divide, and seeking the input of American Indian, Alaska Native, and Native Hawaiian leaders. We will continue to go deep into the Native Nations, meeting collectively and individually with Tribal leaders, Tribal Councils, Native associations, Tribally-owned and operated communications providers, Tribal broadcasters and broadband providers, as well as with Native consumers and businesses.

To obtain a firsthand view of the complexity of the problems, we have been to some of the most unserved areas of the Nation. To see the challenges Native Nations face, we visited some of the most remote schools in the country, such as the Jack Norton School on the Yurok Reservation in California, which is the only school in the state that still operates on a diesel generator.

The school is planned to receive its first ever internet service in a new build out based on an experimental license the Commission granted and one-time federal grant money from the Rural Utilities Service's Community Connect program. We learned more about the important and life-changing impact of broadband when we engaged in distance education discussions from classrooms at the *Native* end of the signals. We learned the true value of high speed internet connections on the island of Moloka'i, where we accepted the gracious invitation of an oncology patient at the *Native* end of the line and sat in on her diagnosis session with her doctors in Honolulu. Hearing the somber diagnosis, like her, we too struggled to read the expressions on the doctors' faces with the lower speed and, therefore, lower resolution connection. In Native Communities, one sees the human element of the lack of communications and broadband services, and the limitations of connectivity, speed, and reliability.

On many occasions we saw impressive solutions juxtaposed with overwhelming great need. For example, on the Cheyenne River Sioux Reservation, we saw the oldest Tribally-owned and operated wireline telecommunications company, the Cheyenne River Sioux Tribe Telephone Authority, deploying fiber to a remote internal valley in their lands. At the Standing Rock Sioux Tribe, we met with their leaders and the management of the Tribe's exciting new wireless company, Standing Rock Telecom, Inc. Two weeks later, we spoke with elected leaders and educators of the Karuk Tribe in the upriver region of the Klamath River in far northern California, who experience little or no wireline or wireless telephone connectivity on their lands. High speed internet is available only at a local computer center. While in Utah some weeks earlier, we met with the leaders of the Confederated Tribes of the Goshute Indian Reservation, who explained that they have been operating for over eight years under a communications stateof-emergency articulated by their Tribal Council--with fcw comprehensive and immediate solutions in sight. Similar examples exist throughout Indian Country and Native Communities.

In addition to our travels to Tribal lands, we have met with many dozen Native Nations and entities at the Commission's headquarters on a myriad of issues involving broadband, broadcast, and telephony matters. On both our travels and in Washington, we have heard many common priorities and concerns.

One such priority is the accurate measurement of the actual state of broadband availability on Tribal lands. Many Tribes have articulated concerns about both the depth and accuracy of the data on the state of services on their lands. Tribal and Native community leaders have asked how this data is verified by the state and federal agencies involved.

In the case of the Goshute Confederated Tribes, during the late September Native American Summit in Salt Lake City, we witnessed their explanation to the Utah state broadband mapping manager that the gross overestimation of the wireless broadband coverage on their reservation actually precluded them from applying for federal grants and loans for a Tribal project that would address the lack of services.

The Utah state broadband mapping coordinator explained that the federal grant did not have funding to verify the data. Increased coordination among the relevant federal agencies and a meaningful involvement of the Native Nations, embracing them as partners, would begin to address these unintended barriers-to-entry.

THE PROCEEDINGS OF MARCH 3, 2011 -- "NATIVE NATIONS DAY" -- NEW COMMISSION APPROACHES

Under the Chairman's leadership, the Commission launched a series of groundbreaking endeavors at its March 3rd Open Meeting, on a day the Commission named "Native Nations Day." It was a day of "firsts"-- the first time that the Commission used its meeting agenda to address matters entirely and specifically developed for Native Nations; the first time that Tribal leaders formally addressed the Commission at the start of an Open Meeting; and the first time that the Commission initiated a comprehensive inquiry and rulemaking proceeding focused exclusively on Native communications needs.

From rules expanding broadcast opportunities, to proposed rules for new mobile wireless licensing opportunities, to an omnibus inquiry on a range of issues related to broadband adoption and deployment on Tribal lands, the proceedings of Native Nations Day will in part serve as the foundation for the nation-to-nation consultation with Native Nations that is a critical component of the Commission's rulemaking process.

THE RURAL RADIO TRIBAL PRIORITY ORDER

Native Nations want to provide information and community news to their people, and are looking at radio programming to promote and preserve Native culture and language, and to advance cultural dialogue. KUYI on the Hopi Reservation, KLND on the Standing Rock Reservation, and KIDE on the Hoopa Valley Reservation are prime examples of such cultural enterprise. Last year, the Commission took steps to address the imbalance in the number of radio stations licensed to Native Nations and communities, as compared to the rest of the country, when it adopted an historic Tribal Priority designed to award a decisive preference to any federally recognized American Indian Tribe or Alaska Native Village seeking to establish its first non-commercial radio station on its Tribal lands. The Tribal Priority was greeted with enthusiasm by Native Nations, but it was noted that certain Native Nations, because of their historical or geographic circumstances, might not be able to take advantage of the priority. In a Second Report and Order adopted on Native Nations Day, the Commission addressed these special circumstances by adopting provisions to address the needs of non-landed Native Nations and those with small or irregularly shaped lands that make it difficult to meet some of the requirements of the Tribal Priority. In addition, the Commission adopted a Notice of Proposed Rulemaking seeking comment on proposals to apply the Tribal Priority to certain commercial FM channel allotments and potentially obviating the need to go to auction. This proceeding is pending at the Commission, and the hope is that these new mechanisms can help Native Nations deploy services in this critical and widely adopted media technology, as they also build designs and resources for new advanced broadband platforms.

THE WIRELESS SPECTRUM TRIBAL LANDS NOTICE OF PROPOSED RULEMAKING

While competitive market forces have spurred robust wireless communications services in many areas of our country, wireless connectivity for Native Nations remains at significantly lower levels. Native Nations have expressed to us many concerns that the situations they face at home involve the very basics of public safety – the inability to make a wireless call in an emergency. Native Nations have asked the Commission for greater access to robust wireless spectrum to meet the challenges of terrain and distance that many Native communities face and, for some time now, the need for this action has been critical. On Native Nations Day, the Commission adopted a Notice of Proposed Rulemaking to promote greater use of spectrum to help close the communications gap on Tribal lands and to ensure that Native Nations are at the center of the decision-making process. This NPRM, one of the most important requests from Native Nations in the last decade, strives to put licenses in the hands of those who will value the spectrum and build out on Tribal lands. This proceeding is pending at the Commission. Three of the five proposals launched in the NPRM would create new opportunities for Native Nations to gain access to spectrum through Commercial Mobile Radio Services licenses, while the other two proposals are designed to create new incentives for existing licensees to deploy wireless services.

THE NATIVE NATIONS NOTICE OF INQUIRY

The Commission has said on many occasions that broadband is indispensable infrastructure for economic growth and job creation, and nowhere is that need more acutely felt than on Tribal lands. The lack of robust broadband services – and, in fact, even basic communications services – contributes to the challenges Native Nations face in building strong economies with diverse businesses and development projects. On Native Nations Day, therefore, the Commission launched a broad-based inquiry into a wide range of communications issues facing Native Nations – an inquiry that will provide a foundation for updating the Commission's rules and policies to provide greater economic, market entry, and communications adoption opportunities and incentives for Native Nations. The result of a broad collaborative effort across the Commission, led by the Office of Native Affairs and Policy, the Notice will lay the groundwork for policies that can help Native Nations build economic and educational opportunities for their own Tribal lands. The Notice seeks comment on the best ways to support sustainable broadband deployment, adoption, and digital literacy training on Tribal lands. Among other important questions, the Commission asks about the possibility of expanding the Tribal Priority concept into a Native Nations Priority, to identify and remove barriers to entry, rather than using a case-by-case waiver approach, thus making it easier for Native Nations to provide other services – wireless, wireline, and satellite – to their communities. The Commission also asks about opportunities to use communications services to help Native Nations address public safety challenges on Tribal lands, including the broad lack of 911 and E-911 services, and the needs of persons with disabilities on Tribal lands.

Recognizing that, given their unique challenges and significant obstacles to broadband deployment, Native Nations need substantially greater financial support than is presently available, the Notice of Inquiry also seeks comment on a recommendation of the National Broadband Plan to establish a Native Nations Broadband Fund.

The National Broadband Plan notes that grants from a new Native Nations Broadband Fund could be used for a variety of purposes, including bringing high-capacity connectivity to governmental headquarters or other anchor institutions, deployment planning, infrastructure build out, feasibility studies, technical assistance, business plan development and implementation, digital literacy, and outreach. In the Notice of Inquiry adopted on Native Nations Day, the Commission seeks comment on a number of issues associated with the establishment of the Native Nations Broadband Fund, including the need for such a fund, the purposes for which it would be used, and the level of funding. The public comment period for the Notice recently ended, and we are in the process of assessing the record and determining next steps for each of the issues addressed in the Notice.

THE LOW-INCOME PROGRAM NOTICE OF PROPOSED RULEMAKING

The Low-Income program of the universal service fund, commonly known as Lifeline and Link Up, has been, and continues to be, a critically important component in extending the reach of communications services to Native Nations. But with a telephone penetration rate hovering below 70 percent and a broadband penetration rate well below ten percent, much remains to be done. According to Gila River Telecommunications, Inc., a Tribally-owned telecommunications company, the telephone penetration rate for the Gila River Indian Community stands at 86 percent, still well below the national average of 98 percent but significantly above the average on Tribal lands. Gila River attributes its success in expanding the reach of telephone service largely to Lifeline, given that roughly 91 percent of the Community's elders participate in Lifeline. At the afternoon session of its March 3rd Open Meeting, the Commission adopted a Notice of Proposed Rulemaking in which it proposes to reform and modernize Lifeline and Link Up – issues of great interest to Native Nations. The Commission is preparing to take action in the near future to address many of the issues raised in the Notice of Proposed Rulemaking.

UNIVERSAL SERVICE REFORM—THE CONNECT AMERICA FUND AND THE MOBILITY FUND

As part of a major rulemaking procedure, the Commission is preparing in the very near future to reform and modernize the High Cost component of the universal service fund, with a proposed transition to a Connect America Fund, including a Mobility Fund. The Office of Native Affairs and Policy is working closely with the Wireline Competition Bureau and the Wireless Telecommunications Bureau to finalize policies that will increase broadband availability— including mobile broadband—in Native Nations, while preserving existing services. In finalizing reforms, we are focused on the unique challenges facing Native Nations, which may not be suitable for a one-size-fits-all solution.

THE FCC-NATIVE NATIONS BROADBAND TASK FORCE

One of the top requests from Native Nations in the National Broadband Plan was the creation of a new FCC-Native Nations Broadband Task Force that would ensure that the Commission's consultation with Native Nations is an ongoing, continuous dialogue and a shared effort between partners. Chairman Genachowski fulfilled this request when, on Native Nations Day, he appointed to the Task Force 19 members representing Native Nations and 11 members representing Bureaus and Offices across the Commission. The Task Force has met twice since its inception – once via conference call and once in person – and is formulating plans to meet again in the near term. The Task Force will ensure that Native concerns are considered in all relevant Commission proceedings and will work to develop additional recommenddations for promoting broadband deployment and adoption on Tribal lands. The Task Force will also coordinate with external entities, including other federal departments and agencies. These efforts will culminate in more efficient ways of working with our Native Nation partners, the industries, and the institutions of Native Nations.

CONCLUSION

The Office of Native Affairs and Policy is ready to continue rolling up our sleeves and pulling out our laptops as we continue our mission. Native Nations Day was a success, and the Commission is proud of the work it has done so far. However, we must build on that success and the success of our other activities since the creation of the Office a mere 14 months ago. Among other things, one of our top priorities is to overhaul, update, and increase the collaborative value of the Commission's Indian Telecom Initiatives, or ITI, program, moving it from version 2.0 to version 10.0 and even beyond. We look forward to increasing the effectiveness and value of these regional workshops, trainings, consultation, and networking events. We also look forward to establishing, by the end of the year, a federal interagency broadband working group that engages other federal agencies concerned with

Native Nations and with missions on Tribal lands related to broadband and communications deployment, such as education, health, public safety, energy, cultural preservation, and economic empowerment. With a new inter-agency initiative on Native broadband, the federal government can coordinate both internally and directly with Native Nations on broadband-related policies and programs.

Internally, we look forward to working with colleagues across the Commission to increase the value of the information tools that the Commission has for Native Nations and Communities. For example, the Commission's Spectrum Dashboard 2.0, which was unveiled in March, allows users to view the licenses and spectrum leases that cover specific or all Tribal lands. We plan to continue holding meetings with Native Nations to discuss how this and other Commission information tools can be improved and more responsive to the needs of Tribal communications planners. We also look forward to reviving an internal training and speaker series for decision makers and colleagues across the Commission on how to work with Native Nations and the basics of how to coordinate and conduct consultations with Native Nations.

In conclusion, we have heard several recurring themes in our conversations with Native leaders – continue to meet with us, listen to us, and use what we tell you to bring communications on Tribal lands into the 21st century. The overarching message is that, if consultations are to be successful, if future education and training sessions are to be well-attended and productive, and if efforts to inform, educate, and put Native Nations at the center of the decision-making process are to succeed, we must do our work with Native Nations largely within their Native communities. Native Nations are aware of our Office's abilities and many have told us that, in order to best help them solve communications problems, we must work with them where the problems exist, see the problems first-hand, and endeavor to find the solutions in concert with them. We welcome all of these opportunities.

Thank you again for the opportunity to testify this afternoon. I look forward to answering any questions you may have.

In: Bridging the Digital Divide in Indian Country ISBN: 978-1-62948-170-8
Editors: R. Louton and J. Sullivan

Chapter 3

STATEMENT OF GEOFFREY BLACKWELL, CHIEF, OFFICE OF NATIVE AFFAIRS AND POLICY, FEDERAL COMMUNICATIONS COMMISSION. HEARING ON "FEDERAL COMMUNICATIONS COMMISSION'S RULE ON THE UNIVERSAL SERVICE FUND AND ITS IMPACT ON AMERICAN INDIANS AND ALASKA NATIVES"*

Chairman Young, Ranking Member Boren, and Members of the Subcommittee, it is an honor to be here before you representing the Federal Communications Commission. Thank you for the opportunity to provide updates on the Commission's progress in reforming the Universal Service Fund and our continued efforts to collaborate with Tribal and Native leaders on those reforms to ensure a sustainable and connected future for Tribal Nations and Native Communities.

Last year, I testified twice before the United States Senate Committee on Indian Affairs concerning the Commission's monumental task to reform an outdated and broken Universal Service Fund.[1] I also provided an overview of the other proceedings where the Commission is engaged on issues to improve

---

* This is an edited, reformatted and augmented version of a statement, Presented June 8, 2012 before the House Committee on Natural Resources, Subcommittee on Indian and Alaska Native Affairs.

the connections to Tribal Nations and Native Communities. I would like to incorporate my earlier testimony to the Senate, due to its extensive coverage of our ongoing work and focus my testimony solely on the topic of concern for today's hearing: the Commission's implementation of USF reform and the progress made since December 2011.

I will first discuss the Commission's implementation of the high-cost fund portion of USF reform. I will then turn to our reform of the low-income programs supported by USF called Lifeline and Link Up, which the Commission adopted in January. In all of these reforms, and in the spirit of the unique trust relationship the Commission shares with Tribal Nations, the Commission has undertaken an unprecedented amount of internal coordination and focus on Tribal issues and has organized extensive consultative outreach efforts and coordination with Tribal Governments, Tribally-owned carriers, and Native Community representatives. As a result, our reforms recognize the unique circumstances and challenges of providing service on Tribal Lands and take steps to directly address these challenges. In summary, among other things, our reforms, for the first time:

- Require all recipients of USF dollars serving Tribal Lands to provide detailed annual reports and to directly and meaningfully engage with Tribal Governments on several key issues critical to the deployment of broadband services on Tribal Lands. Recognizing the important role that all parties play in expediting services to Tribal Lands, these issues include:
    - A needs assessment and deployment planning with a focus on Tribal community anchor institutions;
    - Feasibility and sustainability planning;
    - Marketing services in a culturally sensitive manner;
    - Rights of way processes, land use permitting, facilities siting, environmental and cultural preservation review processes; and
    - Compliance with Tribal business and licensing requirements.
- Target funds directly to locations with unserved Americans, including the over one million unserved consumers on Tribal Lands.
- Dedicate hundreds of millions in support to address the significant mobile broadband gap on Tribal Lands.
- Include specific considerations specifically for Tribally-owned carriers; for example, through bidding credits in the first phase of the Mobility Fund, special Tribal and Alaska variables in the reform of

our legacy rate-of-return mechanisms, and in the waiver process outlined for the Connect America Fund.

- Recognize the unique challenges of serving remote areas of Alaska by including a two-year delay in the reforms to competitive eligible communications carriers in remote Alaska starting on July 1.
- Maintain the Tribal Link Up program while otherwise eliminating Link Up on non-Tribal Lands.
- Expanding the list of Lifeline qualifying programs for Tribal consumers, while recognizing that Tribal consumers should continue to be eligible for enhanced Lifeline support, which is higher than provided on non-Tribal Lands.
- Establishing a Lifeline Broadband Adoption Pilot Program, with the intent to include at least one pilot project on Tribal Lands.

While not an exhaustive list of steps the Commission has taken to address the needs of consumers on Tribal Lands and Tribally-owned carriers, I hope it is clear that the Commission, and I personally, take our statutory responsibility to advance the deployment of voice and broadband service on Tribal Lands very seriously. We also take very seriously the special government-to-government relationship that we share with federally recognized American Indian Tribes and Alaska Native Villages, and exercising our responsibilities with respect to Hawaiian Home Lands.

High-Cost USF Reform—The Connect America Fund

As you are aware, on October 27, 2011 the FCC took a momentous step to reform the high-cost portion of the Universal Service Fund in order to transform it from a program that supports voice telephone service, to one that will also explicitly provide for the deployment of broadband-capable networks, in high-cost areas. The Commission spent significant time and effort developing proposals to best accomplish this. It was a reflection of the fundamental fact that the communications needs of our nation have evolved over time due to technological advancements, and that the Fund desperately needed to be reformed to reflect those significant changes.

Since its inception, the Fund has been instrumental in providing affordable phone service to most Americans; however, there still exists an unacceptable gap in Indian Country. Nevertheless, the Fund has helped achieve an overall higher telephone penetration rate in our nation, which currently stands at 95.6 percent.[2] While that remarkable achievement of the Fund should not be overlooked, technology has progressed rapidly in the last decade. More Americans are using mobile service than ever before, and broadband access is now the gateway by which most Americans obtain critical information and services. As such, it is hard to deny the fact that every American needs access to both voice and broadband services. And while the Commission permitted phone companies to invest USF monies in broadband-capable networks, the Fund's mechanisms were not designed to target support to areas where the private sector will not go, areas where broadband is not being built.

Sadly, our nation remains digitally divided. More than 18 million Americans lack access to robust broadband-capable networks at home. Over a million of those Americans live on Tribal Lands. Indeed, while approximately 6% of all Americans are unserved by broadband, it is well known that Tribal Nations and Native Communities lag well behind the nation as a whole. These Americans are the "have nots" of the broadband world, and they are denied access to some of the most basic features of modern life, that you or I take for granted: optimally navigating your bank account in real time, accessing late breaking news, or quickly sending an email to a friend.

Citizens and communities are significantly disadvantaged without high-speed Internet. Whether you are in urban or rural America, it's hard to find a job, complete an application, operate a small business, or finish your homework without broadband. In fact, 80% of Fortune 500 companies require that you apply online, and high school students with access to broadband at home have higher graduation rates.

For Tribal Nations, having robust broadband networks is even more critical. For Tribal Governments themselves, the benefits of broadband infrastructures will enable new opportunities for the provision of quality healthcare, education, public safety, and jobs. Broadband must be available, accessible, and affordable to meet its great promise for Tribal Nations and Native Communities. Perhaps most importantly, the foundation of robust and viable broadband infrastructure brings Tribal Nations the genuine ability to create opportunities for economic empowerment and the ability to build systems of governance that engage industries, demonstrate stability, and encourage the deployment of services.

FCC Chairman Genachowski crafted a plan that transforms the way $4.5 billion of the high-cost portion of USF is spent each year. The Commission's proceeding to reform USF was unprecedented. The Commission worked with many interested parties, including for example, our sister agency, the Rural Utilities Service at USDA, public interest and consumer advocates, industry, State governments, including State PUCs, and Tribal Governments, among many others. The Commission held workshops both inside and outside the Beltway, and met with numerous parties multiple times. The reform plan drew from many competing ideas, to form a balanced framework that the Commission unanimously approved because it makes the Fund more efficient and effective and will promote significant broadband deployment, as quickly as possible, to millions of unserved consumers in our nation.

The Commission set out five goals for the Fund: (1) to preserve and advance universal availability of voice service; (2) to ensure universal availability of modern networks capable of providing voice and broadband services to homes, businesses, and community anchor institutions; (3) ensure universal availability of modern networks capablc of providing mobile voice and broadband service where Americans live, work , and travel; (4) ensure that rates are reasonably comparable in all regions of the nation, for voice as well as broadband services; and (5) minimize the universal service contribution burden on consumers and businesses.

In Phase I, the reform plan provides for speedy broadband deployment to unserved consumers with a commitment of $650 million in capital in 2012, for both fixed and mobile technologies. While we begin to transition legacy support from those areas that don't need assistance, we can begin to make a difference in those areas that are lacking broadband service. Of that amount, $50 million is reserved the first year alone for mobile service to be extended to Tribal Lands – this is part of the Mobility Fund discussed below. The Commission expects that hundreds of thousands of consumers will be served with fixed broadband networks as a result of Phase I, and tens of thousands of miles of unserved roads will be covered by mobile networks.

The reform plan preserves about $2 billion in funding for those rural carriers that currently operate under the rate-of-return regulatory regime and serve less than five percent of all telephone access lines in the U.S. and have on average reached, according to the rural carrier associations, 92% of their customers with high-speed internet access – that is, service beyond dial up. For these carriers, our reform recognized the unique nature of these providers—most of which are small businesses. As such, rather than making extreme changes, as some interested parties had called for in the proceeding, the Commission focused instead on modifying their current funding mechanisms in a targeted manner to better address our new goals for universal service. For each carrier, the reforms will have a different impact. Some carriers will obtain additional funds, while others will see a decrease in funding. For example, in 2012, we estimate that five of the nine Tribally-owned carriers currently receiving USF support will experience no change or a slight increase in funding, while three of these carriers will see decreases between one and five percent of their total high-cost USF support, and one will see a decrease of less than ten percent.[3] Significantly, our reform carefully balanced the need for certainty and predictability for carriers by avoiding flash cuts and providing transitions so they could adjust to the changes.

In addition, our Order recognized the fact that there could be instances where the Commission will need to specifically review the impact of the total reforms on a particular provider and the consumers it serves through a waiver process. To date, the Commission has received five waiver requests from rate-of-return companies, which are currently under review. The waiver process will give the Commission an opportunity to use a safety net in order to ensure *consumers* aren't inadvertently harmed by our reforms. Of course, given the Commission's role as the steward of the public's money, the waiver process will focus on the financial health of the providers and whether reform jeopardizes their ability to provide service to consumers. The Commission directed the Bureaus to prioritize review of any applications for waiver filed by providers serving Tribal Lands, and to complete the review of those petitions from providers serving Tribal Lands within 45 days of the record closing on such waiver petitions. In addition to the general standard of review, waiver applicants serving Tribal Lands and insular areas are specifically asked to share any additional information about the operating conditions, economic conditions, or other reasons warranting relief based on the unique characteristics of those communities, including information regarding outstanding loans, which the Commission will closely consider when reviewing waiver requests.

In Phase II of the reform, which will begin in 2013, we have allocated approximately $1.8 billion to reach consumers who currently have no broadband service in price cap companies' territories and to maintain existing service. The Commission found that over 80% of the unserved Americans live in the areas served by price cap carriers. As a result, Commission staff is now engaged in a process that will culminate in the adoption of a cost model to determine the amount needed for each carrier to reach those consumers who do not benefit from Phase I. In order to qualify for that money, each price cap carrier will need to make a state-wide commitment to build broadband and serve for five years in its areas where no other provider offers broadband service. Where a commitment is not made, those funds will be available for other providers to offer service and build broadband to unserved consumers through a reverse auction.

Finally, for the first time, the Commission has determined that mobility service should be funded, and the Commission has provided for $500 million of ongoing support through the Mobility Fund, of which $100 million is exclusively for support on Tribal Lands, to ensure coverage of as many roads as possible where mobile carriers will not serve without financial assistance. Additionally, Tribally-owned or controlled entities are eligible for the general, non-Tribal Lands specific Mobility Fund. The Commission also has set aside at least $100 million for the Remote Areas Fund. This will help bring broadband to those Americans who cannot be reached through any of the other high-cost mechanisms.

By phasing out identical support, as many had called for, limited funds may be refocused from supporting multiple networks in a geographic area to ensuring that a broadband-capable network reaches as many Americans as possible. The Commission is shifting money from areas that do not need support (or as much support) to areas where broadband networks are still needed.

With oversight by the FCC, State Commissions, and engagement with Tribal Governments, the Commission also is requiring that providers meet firm deadlines with appropriate accountability and reporting for the money spent. Moreover, by setting a budget the Commission is helping to stabilize the contribution factor and minimize the burden on consumers who and businesses that pay into the fund.

Of great importance is the beneficial impact this reform will have on consumers and their communities, including those living and working on Tribal Nations and in Native Communities. It is a universally-accepted principle that we *all* benefit more when everyone is connected. As such, this reform recognizes that broadband, both fixed and mobile, are the services consumers are demanding and that the Fund should support the networks that offer such services where there is not a private sector business case to be made. And as a condition for receiving high-cost USF assistance, all recipients must deploy broadband-capable networks, in addition to offering voice service to consumers. Accordingly, the Commission expects that millions of consumers will get fixed and mobile broadband coverage where they live, work, and travel as a result of our reform over the next six years.

RECENT STEPS FOR IMPLEMENTING HIGH-COST USF REFORM

There are numerous steps the agency must take to implement reform. For rate-of-return carriers, the Wireline Competition Bureau completed work at the end of April on reasonable limits for reimbursement for high-cost loop support (HCLS). By way of background, HCLS provides close to $800 million annually to help offset high capital and operating expenses faced by rate-of-return carriers. However, under the old system, HCLS lacked benchmarks for determining whether subsidies were warranted under the old system. Because the rules fully subsidized high expense levels, support ran out every year for carriers that maintained lower expenses, effectively punishing efficient operations. The new benchmarks impose reasonable limits on subsidized capital and operating expenses by comparing spending among similarly situated companies. The net effect is that more funding is freed for hundreds of small rural carriers, which in turn will help connect more Americans. Carriers serving 705 study areas across the country will receive more funding for broadband as a result of the benchmarks, and carriers serving 129 study areas that have unusually high expenses will have to take steps to bring their operations more in line with their similarly-situated peers.

In implementing these benchmarks, the agency conducted hundreds of meetings and had ongoing dialogue with and input from all interested parties which led to a significantly improved analysis. For example, the benchmarks now take into account a number of additional factors regarding a carrier's territory, including whether the carrier serves Tribal Lands, and whether it operates in Alaska. In addition, the Bureau modified the timeline so carriers will have more time to adjust to the changes, over an 18-month period. For the Tribally-owned carriers, only one, Mescalero Apache Telecom, Inc., was capped—and its changes will be phased in. The remaining seven out of the eight Tribally-owned carriers receiving HCLS should see increases in their HCLS over the next 18 months.[4] Additionally, the Commission estimates that only two Alaska carriers, Arctic Slope Telephone Association Cooperative and Matanuska Telephone Association, were capped – and their changes will be phased in. The remaining 15 out of 17 rate-of-return carriers in Alaska will see an increase in HCLS.

In addition, the Wireline Competition Bureau announced Connect America Fund Phase I amounts for price cap carriers. CAF Phase I allocates $300 million to extend broadband up to 400,000 currently unserved homes, businesses, and community anchor institutions in rural America, including many unserved locations on Tribal Lands. For example, Alaska Communications Systems has been allocated almost $4.2 million for building broadband to unserved consumers in the State of Alaska, while in Hawaii, Hawaiian Telecom has been allocated just over $400,000 to extend broadband to unserved consumers in its area. In addition, CenturyLink, which serves the most Tribal Lands of any price cap carrier, has been allocated almost $90 million. Price cap carriers have until July 24 to notify the Commission the amounts of money they plan to use, and the census blocks they will commit to build broadband in for receiving CAF Phase I support. While carriers are not required to participate, and we expect some carriers may accept only a portion of the money in light of the rigorous accountability and buildout requirements that accompany it, the Commission anticipates that hundreds of thousands of Americans will gain access to broadband even if carriers only accept a part of the available funds. Whatever amounts are not accepted, the Commission is committed to using the funds to further our broadband goals for the nation. The Commission expects that the expansion of broadband facilities during the CAF Phase I potentially will impact Tribal Lands, and the Commission will keep you apprised of the building commitments we receive from the carriers in and around Tribal Lands.

MOBILITY FUND PHASE I AND PHASE II

Phase I of the Mobility Fund will provide $300 million one-time support, with an additional $50 million allocated to the Tribal Mobility Fund. For Phase II of the Mobility Fund, the Commission proposed to provide recurring support for mobile service, with an annual budget of $500 million, of which up to $100 million will be designated annually for the Tribal Mobility Fund, over a 10-year period. The Commission also sought comment on alternative proposals.

Carriers seeking to serve Tribal Lands may participate in both phases of the general Mobility Fund and the Tribal Mobility Fund. In addition, Tribally-owned or controlled providers seeking general or Tribal Mobility Fund Phase I support for the purpose of providing service on Tribal Lands will receive a 25 percent bidding credit, thus increasing the likelihood that Tribally-owned or controlled entities will receive funding and creating an atmosphere conducive to Tribal economic opportunity and development. Earlier this year, the Wireless Telecommunications Bureau scheduled the Mobility Fund Phase I reverse auction for September 27, 2012. Deemed Auction 901, it will determine (1) which providers will receive Mobility Fund Phase I support to deploy 3G or better wireless service, (2) the specific geographic areas that the providers will cover in exchange for support, and (3) the level(s) of support the providers will receive. Eligible areas will include census blocks unserved today by mobile broadband services, and carriers may not receive support for areas they have previously stated they plan to cover. The auction will maximize coverage of unserved road miles within the budget, and winners will be required to deploy 4G service within three years, or 3G service within two years, accelerating the migration to 4G. Mobility Fund recipients will be subject to public interest obligations, including data roaming and collocation requirements.

The Wireless Bureau completed the final auction procedures in May, including finalizing the list of census blocks eligible to receive support in the auction. Over 50 interested parties commented on the procedures and/or census block eligibility. The Bureau anticipates rolling out a tutorial for prospective bidders prior to opening the short-form filing window on June 27th. In order to participate in Auction 901, a prospective applicant must file a short-form application by July 11, 2012. After the auction, each winning bidder will be required to submit a long-form application to qualify for Phase I support. The Commission anticipates making the first disbursement of Phase I support by early 2013.

The Bureau is also currently considering petitions by three carriers for designation as eligible telecommunications carriers (ETCs) in order to be eligible to receive USF support and participate in the Mobility Fund Phase I auction. In addition, the Bureau has received information from State Commissions about new ETC designations for carriers interested in receiving Mobility Fund support.

CONNECT AMERICA FUND PHASE II

Staff also has begun the crucial work for moving to Phase II of our reforms. I will refer you to a recent report released on June 1st that describes the agency's efforts to develop the cost model for the price cap territories for Phase II.[5] Last week, Commission staff also initiated a process to gather data regarding wire center boundaries for price cap carriers, which is necessary to determine where support should be targeted. This week, Commission staff will be soliciting public input on the key questions that need to be resolved in order to develop such a model.

The Commission has routinely solicited public comment in its modernization of the Fund. The Connect America Fund docket has been filled with an unprecedented amount of public comments on proposals offered by the Commission and others. The Commission has been working very hard to hear all parties' concerns and address them in a timely fashion. For example, the Commission has received numerous Petitions for Reconsideration and/or Clarification, and the Commission has already addressed numerous issues in three separate Commission level decisions. Moreover, Commission staff has issued multiple Orders that have offered clarifications that have been requested, and have spent hours on conference calls with State Commission staff and other parties, including several visits to Tribal Lands specifically focused on USF reform since the order was released, answering questions about implementation details. At all times, the Commission has had an open door policy to hear from all interested parties, and the Commission has worked diligently towards resolutions that benefit consumers and help us achieve our universal service goals.

Lifeline and Link Up Reform

At the end of January, the Commission comprehensively reformed and modernized the Lifeline and Link Up programs. For the past 25 years, Lifeline helped tens of millions of low-income Americans afford basic phone service, and the percentage of low-income households with phone service has increased from 80% in 1985, when Lifeline began, to nearly 92% last year. However, the program was facing significant challenges that needed to be addressed, including the fact that consumers now have numerous providers from which to choose as compared to when Lifeline first began. This had the inadvertent result of some consumers obtaining more than one Lifeline benefit. In addition to addressing this issue by creating a National Lifeline Accountability Database to prevent multiple carriers from receiving support for the same subscriber, the Commission also made a number of other changes to ensure that the Lifeline program is as efficient and effective as possible, while meeting the Commission's newly instituted goals for the program to: (1) ensure the availability of voice service for low-income Americans; (2) ensure the availability of broadband service for low-income Americans; and (3) minimize the contribution burden on consumers and businesses.

Lifeline is also extremely important to consumers on Tribal Lands and in Native Communities. Some of these consumers face crushing cyclical poverty in Indian Country. Some Tribal Nations have poverty rates over 70%, such as the Standing Rock Sioux Tribe, which owns and operates Standing Rock Telecom, Inc, the most recently designated Tribal ETC. Through Lifeline reform, the Commission has made its commitment to connecting these communities a priority. The Commission worked to ensure that eligible families who need phone service can participate in the program in a way that makes sense for them – be it mobile or wireline service. The Commission addressed concerns about waste, fraud, and abuse through rigorous but not burdensome new certification and verification requirements. Importantly, the Commission ensured that the needs of Tribal communities would continue to be addressed as Universal Service transforms toward broadband. In the past, the Commission acknowledged the low subscribership to telecommunications services and low-income needs on Tribal Lands and provided enhanced Lifeline and Link Up support to address those unique challenges. For example, in 2000, only an estimated 47 percent of Tribal households had phone service compared to 94 percent of all American households.

In reforming the Lifeline program, the Commission set a uniform Lifeline discount amount at $9.25 per month on non-Tribal Lands, but recognized that low-income consumers living on Tribal Lands should continue to be eligible for enhanced Lifeline support—up to an additional $25 per month in Lifeline support. The Commission eliminated Link Up support (which off-sets the cost of commencing service) on non-Tribal Lands due to the evidence that Link Up was no longer needed, but maintained the enhanced Link Up program on Tribal Lands for carriers receiving high-cost support, due to the significant telecommunications and connectivity challenges on Tribal Lands. As a result, to initiate phone service on Tribal Lands for qualifying low-income consumers, Link Up provides up to $100.

The Commission also took steps to further advance the availability of Lifeline support for low-income consumers living on or near Tribal Lands. First, the Commission amended its rules to clarify that low-income residents of Tribal Lands may be eligible for program support based on either income or participation in certain federal or Tribal assistance programs. Second, the Commission amended its rules to expand program-based eligibility to participants in the Food Distribution Program on Indian Reservations (FDPIR), a federal program that provides food to low-income households living on Indian reservations and to Native American families residing in designated areas near reservations and in the State of Oklahoma. Third, the

Commission established a process for Tribal Governments to seek designation of off-reservation Lands as Tribal Lands for the purpose of receiving enhanced Lifeline support, and clarified that low-income residents of Tribal Lands may self-certify as to their residency on Tribal Lands to receive additional support. The effects of this process will be that those who otherwise qualify for enhanced Lifeline support be deemed eligible for such while giving Tribal Governments the tools necessary to ensure that their otherwise eligible Tribal members are able to obtain enhanced Lifeline service.

The Lifeline and Link Up support received by those carriers serving Tribal Lands can be a significant portion of their total USF support and further supports the business case for carriers on Tribal Lands. For example, in 2010, $102.7 million was provided to carriers to reimburse them for providing phone service to low-income households on Tribal Lands.[6] Thus, the low-income programs not only ensure that phone service is available for consumers, they also support the carriers' ongoing ability to offer service in their territories.

The Commission also recognizes that the Lifeline program can play an important role in helping to close the broadband adoption gap, while improving the business case for carriers to serve communities in Tribal Lands. As part of its plan to modernize the Lifeline program, the Commission recently launched a Broadband Pilot Program and will be accepting applications on July 2nd. The Commission is very interested in receiving applications for pilot projects on Tribal Lands, and provided some additional flexibility for Tribally-owned carriers to participate, and the Commission looks forward to continuing our work with Tribal Nations and Native Communities to learn more about what drives broadband adoption.

Continued Engagement with Tribal Nations and Native Communities

As the Commission implements its reforms, the Office of Native Affairs and Policy (ONAP) continues to engage with Tribal Nations and Native Communities to explain the changes, hear from them about what adjustments the Commission may need to make, and coordinate with them on the opportunities that these reforms offer them. This represents the unprecedented level of engagement and analysis that Tribal Nations and Native Communities are now experiencing at the Commission.

In this High-Cost Reform Connect America Fund proceeding alone, dozens of Tribal specific questions were presented in the proposed rulemaking, and the Commission engaged Tribal Governments and associations, Native institutions and community leaders, throughout the nation. We established an open and direct dialogue through ONAP and the Bureaus. On two separate occasions, both in the spring and fall of 2011, the National Tribal Telecommunications Association, the association of Tribally-owned telephone companies, had rounds of meetings with the Chairman and each of the Commissioners. They were joined by two major national inter-Tribal government organizations, the National Congress of American Indians and Affiliated Tribes of NW Indians. The Bureaus and Offices met several times with the individual Tribes and Tribal Nations and companies. The Office of Native Affairs and Policy took its very first beyond the beltway trip as a new office in late 2010 to a quarterly meeting of the Tribally owned telcos, and has consistently worked these regulatory priorities into virtually every effort they undertake on the ground in Indian Country.

ONAP also just recently launched, in May, the next generation of Tribal Nation and Native Community field training programs. These new "Tribal communications 101" meetings and consultations respond to the specific interests and needs of key regional partner organizations. Focused on delivering granular, specific information to Tribal leaders and their staff at the Tribal Planner and IT Director level, these regional training workshops have proven to be exceptionally well-received. Along with wireline, wireless, and broadcast opportunities, the Universal Service Fund reform and the new programs and opportunities for Tribal Nations and Native Communities have been a key feature in this training, and will continue to be. As part of these sessions, our new FCC Native Learning Lab provides hands-on instruction in a "internet café classroom" setting to effectively teach the Commission's web-based systems and applications thereby enhancing coordination with the FCC, improving information flow to Tribal Nation and Native Community decision makers, and ultimately increasing the quality of our consultation with Tribal Nations and the involvement of Native voices within our dockets.

A positive outcome of these regional training sessions has been the development of formal and informal Tribal working coalitions, where leaders and managers from Tribal Nations in varying states of telecom and broadband deployment work together, learning from each others' experiences and mentoring others just starting the journey within their area. Our first two of these targeted meetings were held on the Swinomish Indian Reservation in Washington with the Affiliated Tribes of NW Indians, and on the Pala Indian

Reservation in California with the Southern California Tribal Chairmen's Association. Several other regions and partners are targeted. Our new regional, scalable training workshops will grow and go to other regions throughout the nation, and aim to make a tangible difference in helping Tribal Government and Native Community officials work with the Commission, interact with the communications industries, and lead their people to the myriad benefits of new communications capabilities. The Commission has much work to do, and is committed to meeting the modern communications needs of Tribal Nations and Native Communities.

CONCLUSION

Broadband is one of this generation's most important challenges because it presents one of our most monumental opportunities. A consequence of the unfortunate history that Indian Country has endured is an endemic lack of many critical infrastructures in Tribal Nations and Native Communities. Virtually no critical infrastructure has come to Tribal Lands without federal investment, oversight, and regulation. The potential of broadband and communications technologies can do a great deal to level that terrible history. Extending broadband will help us achieve a more equal opportunity for *all* Americans. If the Commission continues to work together with Tribal Nations and Native Communities to complete and perfect these reform efforts, we can ensure that the transition of the Universal Service Fund from voice to broadband opens the door for every citizen to become a part of our digital future of our country and that Tribal Nations and Native Communities enjoy a secure and enduring place in that future.

Thank you for the opportunity to testify today, and I look forward to addressing your questions.

End Notes

[1] Written Statement of Geoffrey C. Blackwell, Chief, Office of Native Affairs and Policy, Federal Communications Commission, Before the United States Senate Committee on Indian Affairs, "Deficit Reduction and Job Creation: Regulatory Reform in Indian Country," Dec. 1, 2011; Written Statement of Geoffrey C. Blackwell, Chief, Office of Native Affairs and Policy, Federal Communications Commission, Before the United States Senate Committee on Indian Affairs, "Internet Infrastructure in Native Communities: Equal Access to E-Commerce, Jobs and the Global Marketplace," Oct. 6, 2011.

[2] federal communications commission,wireline competition bureau, industry analysis and technology division, telephone subscribership in the united states at 2 (Dec. 2011).

[3] Our staff's projections (based on the carrier's 2010 cost data) indicate that support amounts will be similar for 2013 and 2014.

[4] Carriers are now required to invest their USF support in broadband-capable networks.

[5] Wireline Competition Bureau Issues Progress Report On The Connect America Fund (CAF) Phase II Model, Public Notice (rel. June 1, 2012).

[6] 2011monitoring report at Table 2.2.

In: Bridging the Digital Divide in Indian Country ISBN: 978-1-62948-170-8
Editors: R. Louton and J. Sullivan

Chapter 4

BROADBAND INTERNET ACCESS AND THE DIGITAL DIVIDE: FEDERAL ASSISTANCE PROGRAMS*

Lennard G. Kruger and Angele A. Gilroy

SUMMARY

The "digital divide" is a term that has been used to characterize a gap between "information haves and have-nots," or in other words, between those Americans who use or have access to telecommunications and information technologies and those who do not. One important subset of the digital divide debate concerns high-speed Internet access and advanced telecommunications services, also known as broadband. Broadband is provided by a series of technologies (e.g., cable, telephone wire, fiber, satellite, wireless) that give users the ability to send and receive data at volumes and speeds far greater than traditional "dial-up" Internet access over telephone lines.

Broadband technologies are currently being deployed primarily by the private sector throughout the United States. While the numbers of new broadband subscribers continue to grow, studies and data suggest that the rate of broadband deployment in urban/suburban and high income areas is outpacing deployment in rural and low-income areas. Some

---

* This is an edited, reformatted and augmented version of a Congressional Research Service publication, CRS Report for Congress RL30719, prepared for Members and Committees of Congress, from www.crs.gov, dated July 17, 2013.

policymakers, believing that disparities in broadband access across American society could have adverse economic and social consequences on those left behind, assert that the federal government should play a more active role to avoid a "digital divide" in broadband access.

With the conclusion of the grant and loan awards established by the American Recovery and Reinvestment Act of 2009 (P.L. 111-5), there remain two ongoing federal vehicles which direct federal money to fund broadband infrastructure: the broadband and telecommunications programs at the Rural Utilities Service (RUS) of the U.S. Department of Agriculture and the Universal Service Fund (USF) programs under the Federal Communications Commission (FCC). Although the USF's High Cost Program does not explicitly fund broadband infrastructure, subsidies are used, in many cases, to upgrade existing telephone networks so that they are capable of delivering high-speed services. Additionally, subsidies provided by USF's Schools and Libraries Program and Rural Health Care Program are used for a variety of telecommunications services, including broadband access. Currently the USF is undergoing a major transition to the Connect America Fund, which is targeted to the deployment, adoption, and utilization of both fixed and mobile broadband.

To the extent that the 113th Congress may consider various options for further encouraging broadband deployment and adoption, a key issue is how to strike a balance between providing federal assistance for unserved and underserved areas where the private sector may not be providing acceptable levels of broadband service, while at the same time minimizing any deleterious effects that government intervention in the marketplace may have on competition and private sector investment.

INTRODUCTION

The "digital divide" is a term used to describe a perceived gap between "information haves and have-nots," or in other words, between those Americans who use or have access to telecommunications and information technologies and those who do not.[1] Whether or not individuals or communities fall into the "information haves" category depends on a number of factors, ranging from the presence of computers in the home, to training and education, to the availability of affordable Internet access.

Broadband technologies are currently being deployed primarily by the private sector throughout the United States. While the numbers of new broadband subscribers continue to grow, studies and data suggest that the rate of broadband deployment in urban/suburban and high income areas is outpacing deployment in rural and low-income areas.

STATUS OF BROADBAND DEPLOYMENT IN THE UNITED STATES

Prior to the late 1990s, American homes accessed the Internet at maximum speeds of 56 kilobits per second by dialing up an Internet Service Provider (such as AOL) over the same copper telephone line used for traditional voice service. A relatively small number of businesses and institutions used broadband or high speed connections through the installation of special "dedicated lines" typically provided by their local telephone company. Starting in the late 1990s, cable television companies began offering cable modem broadband service to homes and businesses. This was accompanied by telephone companies beginning to offer DSL service (broadband over existing copper telephone wireline). Growth has been steep, rising from 2.8 million high speed lines reported as of December 1999, to 243 million connections as of June 30, 2012.[2] Of the 243 million high speed connections reported by the FCC, 197 million serve residential users.[3]

Table 1 depicts the relative deployment of different types of broadband technologies. A distinction is often made between "current generation" and "next generation" broadband (commonly referred to as next generation networks or NGN). "Current generation" typically refers to currently deployed cable, DSL, and many wireless systems, while "next generation" refers to dramatically faster download and upload speeds offered by fiber technologies and also by successive generations of cable, DSL, and wireless technologies.[4] In general, the greater the download and upload speeds offered by a broadband connection, the more sophisticated (and potentially valuable) the application that is enabled.

Based on the latest FCC broadband connection data, *Table 2* shows the percentages of households with broadband connections by state, both for download connections over 200 kbps and for connections of at least 3 Mbps (which approximates the FCC's National Broadband Availability target). According to the FCC, high speed connections over 200 kbps are reported in 67% of households nationwide, while connections of at least 3 Mbps (download) and 768 kbps (upload) are reported in 38% of households nationwide.

Table 1. Percentage of Broadband Technologies by Types of Connection

	Connections over 200 kbps in at least one direction	Residential connections over 200 kbps in at least one direction	Connections at least 3 Mbps downstream and 768 kbps upstream	Residential connections at least 3 Mbps downstream and 768 kbps upstream
cable modem	20.4%	23.9%	37.7%	45.3%
DSL	12.8%	14.1%	12.8%	14.7%
Mobile wireless	63.0%	58.2%	42.9%	32.5%
Fiber	2.6%	3.0%	6.0%	7.2%
All other	1.2%	0.8%	0.6%	0.4%

Source: FCC, Internet Access Services: Status as of June 30, 2012, pp. 23-26.

Table 2. Percentage of Households With Broadband Connections by State (as of June 30, 2012)

	Connections over 200 kbps	Connections at least 3 mbps downstream and 768 kbps upstream
Alabama	57%	30%
Alaska	71%	34%
Arizona	69%	54%
Arkansas	54%	25%
California	76%	44%
Colorado	76%	61%
Connecticut	79%	58%
Delaware	78%	69%
District of Columbia	72%	65%
Florida	74%	51%
Georgia	65%	42%
Hawaii	*	*
Idaho	61%	30%
Illinois	67%	44%
Indiana	64%	41%
Iowa	66%	25%
Kansas	67%	33%
Kentucky	62%	38%
Louisiana	60%	30%
Maine	75%	26%
Maryland	76%	70%
Massachusetts	81%	75%
Michigan	65%	46%

	Connections over 200 kbps	Connections at least 3 mbps downstream and 768 kbps upstream
Minnesota	69%	48%
Mississippi	50%	21%
Missouri	62%	29%
Montana	64%	38%
Nebraska	68%	45%
Nevada	66%	50%
New Hampshire	81%	65%
New Jersey	82%	76%
New Mexico	59%	37%
New York	76%	54%
North Carolina	69%	17%
North Dakota	70%	41%
Ohio	68%	25%
Oklahoma	58%	28%
Oregon	69%	55%
Pennsylvania	71%	58%
Rhode Island	76%	*
South Carolina	64%	29%
South Dakota	67%	47%
Tennessee	58%	39%
Texas	65%	33%
Utah	73%	56%
Vermont	77%	61%
Virginia	70%	60%
Washington	72%	61%
West Virginia	60%	49%
Wisconsin	67%	30%
Wyoming	66%	44%
National subscribership ratio	69%	45%

Source: FCC, Internet Access Services: Status as of June 30, 2012, pp. 34-35.

Notes: Asterisk (*) indicates data withheld by the FCC to maintain firm confidentiality. Subscribership ratio is the number of reported residential high speed lines (broadband connections) divided by the number of households in each state.

Meanwhile, the National Broadband Map, which is composed of state broadband data and compiled by NTIA, provides data on where broadband is and is not available. The latest update of these data indicate that 98.2% of the U.S. population have available advertised speeds of at least 3 Mbps (download) and 768 kbps (upload), 96.2% have download speeds 6 Mbps or greater, and 94.4% have download speeds 10 Mbps or greater.[5] The FCC's *Eighth Broadband Progress Report*, released on August 21, 2012, used National Broadband Map data to estimate that 19 million Americans living in 7 million households lack access to fixed broadband at speeds of 4 Mbps (download)/1 Mbps (upload) or greater.[6] *Table 3* shows a state-by-state breakdown of the percentage of population without access to fixed broadband at the FCC's benchmark speed of 4 Mbps/1Mbps.

Table 3. Americans Without Access to Fixed Broadband by State (access to speeds of at least 4 Mbps download/1 Mbps upload)

	% of population without access	% of population without access, nonrural areas	% of population without access, rural areas
United States	6.0%	1.8%	23.7%
Alabama	11.4%	1.6%	25.5%
Alaska	19.6%	4.4%	48.9%
Arizona	4.7%	1.2%	35.8%
Arkansas	13.6%	1.8%	28.8%
California	3.3%	1.6%	35.2%
Colorado	4.3%	1.0%	25.3%
Connecticut	0.7%	0.5%	2.6%
Delaware	3.1%	1.1%	13.0%
District of Columbia	0.0%	0.0%	N/A
Florida	3.1%	2.0%	14.3%
Georgia	3.4%	1.3%	9.9%
Hawaii	1.5%	0.1%	17.7%
Idaho	13.1%	1.3%	41.4%
Illinois	3.3%	0.4%	25.6%
Indiana	4.3%	1.3%	12.4%
Iowa	7.1%	0.7%	18.7%
Kansas	7.7%	1.0%	27.0%
Kentucky	10.5%	1.5%	23.0%
Louisiana	8.8%	1.3%	29.6%
Maine	4.7%	1.2%	7.0%
Maryland	3.2%	0.9%	19.2%
Massachusetts	1.0%	0.5%	6.4%
Michigan	6.3%	0.8%	22.4%
Minnesota	8.0%	0.8%	27.7%
Mississippi	12.1%	1.2%	22.8%
Missouri	7.5%	0.6%	24.2%
Montana	26.7%	4.0%	55.4%

	% of population without access	% of population without access, nonrural areas	% of population without access, rural areas
Nebraska	10.1%	1.9%	33.0%
Nevada	2.3%	0.6%	30.2%
New Hampshire	7.5%	2.5%	15.2%
New Jersey	0.7%	0.4%	5.6%
New Mexico	14.2%	4.8%	46.7%
New York	1.3%	0.0%	10.4%
North Carolina	6.4%	2.1%	15.0%
North Dakota	15.9%	2.5%	36.2%
Ohio	3.4%	0.5%	14.0%
Oklahoma	16.2%	2.9%	42.5%
Oregon	3.4%	0.2%	17.3%
Pennsylvania	1.7%	0.3%	6.8%
Rhode Island	0.2%	0.0%	2.3%
South Carolina	11.7%	4.9%	25.1%
South Dakota	21.1%	3.2%	44.6%
Tennessee	6.8%	0.9%	18.6%
Texas	5.9%	2.0%	27.6%
Utah	1.8%	0.3%	16.7%
Vermont	9.4%	0.2%	15.2%
Virginia	10.9%	2.2%	37.6%
Washington	3.2%	0.5%	17.4%
West Virginia	45.9%	31.4%	59.8%
Wisconsin	6.9%	0.1%	23.0%
Wyoming	13.2%	1.1%	35.4%
U.S. Territories	54.0%	41.5%	85.2%
American Samoa	78.6%	30.9%	92.0%
Northern Marianas	100.0%	100.0%	100.0%
Guam	54.3%	0.1%	76.1%
Puerto Rico	51.6%	40.3%	84.8%
U.S. Virgin Islands	100.0%	100.0%	100.0%

Source: FCC, Eighth Broadband Progress Report, Appendix C.

In contrast to broadband *availability*, which refers to whether or not broadband service is offered, broadband *adoption* refers to the extent to which American households actually subscribe to and use broadband. The U.S. Department of Commerce, based on October 2012 survey data, found that 72.4% of U.S. households have adopted broadband.[7] Similarly, the FCC's *Eighth Broadband Progress Report* found that 64% of American households with broadband available to them adopt broadband service offering speeds faster than 768 kbps/200 kbps, while 40% adopt speeds faster than the FCC benchmark of 4 Mbps/1Mbps. The FCC found that the "broadband adoption rates for American households are lower, on average, in the counties with the lowest median household income, in areas outside of urban areas, on Tribal lands, and in U.S. Territories."[8]

According to the Department of Commerce report *Exploring the Digital Nation: America's Emerging Online Experience* (based on July 2011 U.S. Census Bureau survey data), the three main reasons cited for not having broadband at home are that it is perceived as not needed, too expensive, and lack of a home computer.[9] The Department of Commerce report, the FCC's National Broadband Plan, and a survey conducted by the Pew Internet and American Life Project[10] also found disparities in broadband adoption among demographic groups. Populations continuing to lag behind in broadband adoption include people with low incomes, seniors, minorities, the less-educated, non-family households, and the non-employed.

BROADBAND IN RURAL AREAS[11]

While the number of new broadband subscribers continues to grow, the rate of broadband deployment in urban areas appears to be outpacing deployment in rural areas.

While there are many examples of rural communities with state of the art telecommunications facilities,[12] recent surveys and studies have indicated that, in general, rural areas tend to lag behind urban and suburban areas in broadband deployment. For example:

- According to the FCC's *Eighth Broadband Progress Report*, released in August 2012, of the 19 million Americans who live where fixed broadband is unavailable, 14.5 million live in rural areas.[13]
- According to June 2012 data from the National Broadband Map, 99.6% of the population in urban areas have access to available broadband download speeds of at least 6 Mbps, as opposed to 81.8% of the population in rural areas.[14]
- The Department of Commerce report, *Exploring the Digital Nation: America's Emerging Online Experience*, found that while the digital divide between urban and rural areas has lessened since 2007, it still

persists with 72% of urban households adopting broadband service in 2011, compared to 58% of rural households.[15]

The comparatively lower population density of rural areas is likely the major reason why broadband is less deployed than in more highly populated suburban and urban areas. Particularly for wireline broadband technologies—such as cable modem and DSL—the greater the geographical distances among customers, the larger the cost to serve those customers. Thus, there is often less incentive for companies to invest in broadband in rural areas than, for example, in an urban area where there is more demand (more customers with perhaps higher incomes) and less cost to wire the market area.[16]

Some policymakers believe that disparities in broadband access across American society could have adverse consequences on those left behind, and that advanced telecommunications applications critical for businesses and consumers to engage in e-commerce are increasingly dependent on high speed broadband connections to the Internet. Thus, some say, communities and individuals without access to broadband could be at risk to the extent that connectivity becomes a critical factor in determining future economic development and prosperity. A February 2006 study done by the Massachusetts Institute of Technology for the Economic Development Administration of the Department of Commerce marked the first attempt to quantitatively measure the impact of broadband on economic growth. The study found that "between 1998 and 2002, communities in which mass-market broadband was available by December 1999 experienced more rapid growth in employment, the number of businesses overall, and businesses in IT-intensive sectors, relative to comparable communities without broadband at that time."[17]

A June 2007 report from the Brookings Institution found that for every one percentage point increase in broadband penetration in a state, employment is projected to increase by 0.2% to 0.3% per year. For the entire U.S. private non-farm economy, the study projected an increase of about 300,000 jobs.[18]

Subsequently, a July 2009 study commissioned by the Internet Innovation Alliance found net consumer benefits of home broadband on the order of $32 billion per year, up from an estimated $20 billion in consumer benefits from home broadband in 2005.[19]

Some also argue that broadband is an important contributor to U.S. future economic strength with respect to the rest of the world. Data from the Organization for Economic Cooperation and Development (OECD) found the U.S. ranking 15th among OECD nations in broadband access per 100 inhabitants as of December 2011.[20] By contrast, in 2001 an OECD study found the U.S. ranking fourth in broadband subscribership per 100 inhabitants (after Korea, Sweden, and Canada).[21] While many argue that declining U.S. performance in international broadband rankings is a cause for concern,[22] others maintain that the OECD data undercount U.S. broadband deployment,[23] and that cross-country broadband deployment comparisons are not necessarily meaningful and are inherently problematic.[24] Finally, an issue related to international broadband rankings is the extent to which broadband speeds and prices differ between the United States and the rest of the world.[25]

ARE BROADBAND DEPLOYMENT DATA ADEQUATE?

Obtaining an accurate snapshot of the status of broadband deployment is problematic. Anecdotes abound of rural and low-income areas which do not have adequate Internet access, as well as those which are receiving access to high-speed, state-of-the-art connections. Rapidly evolving technologies, the constant flux of the telecommunications industry, the uncertainty of consumer wants and needs, and the sheer diversity and size of the nation's economy and geography make the status of broadband deployment very difficult to characterize. The FCC periodically collects broadband deployment data from the private sector via "FCC Form 477"—a standardized information gathering survey. Statistics derived from the Form 477 survey are published every six months. Additionally, data from Form 477 are used in the FCC's (to date) eight broadband deployment reports.

The FCC is working to refine the data used in future reports in order to provide an increasingly accurate portrayal. In its March 17, 2004, Notice of Inquiry for the *Fourth Report*, the FCC sought comments on specific proposals to improve the FCC Form 477 data gathering program.[26] On November 9, 2004, the FCC voted to expand its data collection program by requiring reports from all facilities based carriers regardless of size in order to better track rural and underserved markets, by requiring broadband providers to provide more information on the speed and nature of their service, and by establishing broadband-over-power line as a separate category in order to track its development and deployment. The FCC Form 477 data gathering program was extended for five years beyond its then March 2005 expiration date.[27]

On April 16, 2007, the FCC announced a Notice of Proposed Rulemaking which sought comment on a number of broadband data collection issues, including how to develop a more accurate picture of broadband deployment; gathering information on price, other factors determining consumer uptake of broadband, and international comparisons; how to improve data on wireless broadband; how to collect information on subscribership to voice over Internet Protocol service (VoIP); and whether to modify collection of speed tier information.[28]

On March 19, 2008, the FCC adopted an order that substantially expands its broadband data collection capability. Specifically, the order expands the number of broadband reporting speed tiers to capture more information about upload and download speeds offered in the marketplace, requires broadband providers to report numbers of broadband subscribers by census tract, and improves the accuracy of information collected on mobile wireless broadband deployment.

Additionally, in a Further Notice of Proposed Rulemaking, the FCC sought comment on broadband service pricing and availability.[29] The July 2009 data release (providing data as of June 30, 2008) was the final data set gathered under the old FCC Form 477. The February 2010 data report (December 31, 2008, data) reflected the new Form 477 data collection requirements. Meanwhile, during the 110th Congress, state initiatives to collect broadband deployment data in order to promote broadband in underserved areas were viewed as a possible model for governmental efforts to encourage broadband.

The Broadband Data Improvement Act was enacted by the 110th Congress and became P.L. 110-385 on October 10, 2008. The law requires the FCC to collect demographic information on unserved areas, data comparing broadband service with 75 communities in at least 25 nations abroad, and data on consumer use of broadband. The act also directs the Census Bureau to collect broadband data, the Government Accountability Office to study broadband data metrics and standards, and the Department of Commerce to provide grants supporting state broadband initiatives.

P.L. 111-5, the American Recovery and Reinvestment Act, provided NTIA with an appropriation of $350 million to implement the Broadband Data Improvement Act and to develop and maintain a national broadband inventory map. The National Broadband Map was first released on February 17, 2011 (http://www.broadbandmap.gov), and is updated every six months.[30]

Finally, the FCC's National Broadband Plan addressed the broadband data issue, recommending that the FCC and the U.S. Bureau of Labor Statistics (BLS) should collect more detailed and accurate data on actual availability, penetration, prices, churn, and bundles offered by broadband service providers to consumers and businesses, and should publish analyses of these data.

BROADBAND AND THE FEDERAL ROLE

The Telecommunications Act of 1996 (P.L. 104-104) addressed the issue of whether the federal government should intervene to prevent a "digital divide" in broadband access. Section 706 requires the FCC to determine whether "advanced telecommunications capability [i.e., broadband or high-speed access] is being deployed to all Americans in a reasonable and timely fashion."

Since 1999, the FCC has adopted and released eight reports pursuant to Section 706. The first five reports formally concluded that the deployment of advanced telecommunications capability to all Americans is reasonable and timely. Unlike the first five 706 reports, the sixth, seventh, and eighth reports concluded that broadband is not being deployed to all Americans in a reasonable and timely fashion. According to the *Eighth Broadband Progress Report*:

> Our analysis shows that the nation's broadband deployment gap remains significant and is particularly pronounced for Americans living in rural areas and on Tribal lands. We find that as of June 30, 2011, approximately 19 million Americans did not have access to fixed broadband. Significantly, approximately 76 percent of these Americans reside in rural areas. Our analysis further shows that Americans residing on Tribal lands disproportionately lack access to fixed broadband. And the available international broadband data, though not perfectly comparable to U.S. data, suggest that the availability and deployment of broadband in the United States may lag behind a number of other developed countries in certain respects, although we also compare favorably to some developed countries in other respects. Moreover, as many as 80 percent of E-rate recipients say that their broadband connections do not fully meet their needs, and 78 percent of recipients say that they need additional bandwidth. These data combined with our findings concerning availability above provide further indication that broadband is not yet being reasonably and timely deployed to all Americans.[31]

FCC Commissioners Robert McDowell and Ajit Pai issued dissenting statements, maintaining that there is insufficient justification for the 706 report conclusion that broadband is not being deployed in a reasonable and timely fashion. For example, the dissents argued that the report did not sufficiently account for the dramatic growth in the availability and deployment of mobile broadband, and that gaps in broadband adoption should not be used to determine whether or not broadband is being sufficiently deployed.[32]

The National Broadband Plan

As mandated by the ARRA, on March 16, 2010, the FCC publically released its report, *Connecting America: The National Broadband Plan.*[33] The National Broadband Plan (NBP) seeks to "create a high-performance America," which the FCC defines as "a more productive, creative, efficient America in which affordable broadband is available everywhere and everyone has the means and skills to use valuable broadband applications."[34] In order to achieve this mission, the NBP recommends that the country set six goals for 2020:

- Goal No. 1: At least 100 million U.S. homes should have affordable access to actual download speeds of at least 100 megabits per second and actual upload speeds of at least 50 megabits per second.
- Goal No. 2: The United States should lead the world in mobile innovation, with the fastest and most extensive wireless networks of any nation.
- Goal No. 3: Every American should have affordable access to robust broadband service, and the means and skills to subscribe if they so choose.
- Goal No. 4: Every American community should have affordable access to at least 1 gigabit per second broadband service to anchor institutions such as schools, hospitals, and government buildings.
- Goal No. 5: To ensure the safety of the American people, every first responder should have access to a nationwide, wireless, interoperable broadband public safety network.
- Goal No. 6: To ensure that America leads in the clean energy economy, every American should be able to use broadband to track and manage their real-time energy consumption.

The National Broadband Plan is categorized into three parts:

- *Part I (Innovation and Investment),* which "discusses recommend-dations to maximize innovation, investment and consumer welfare, primarily through competition. It then recommends more efficient allocation and management of assets government controls or influences."[35] The recommendations address a number of issues, including spectrum policy, improved broadband data collection, broadband performance standards and disclosure, special access rates, interconnection, privacy and cybersecurity, child online safety, poles and rightsof-way, research and experimentation (R&E) tax credits, and R&D funding.

- *Part II (Inclusion),* which "makes recommendations to promote inclusion—to ensure that all Americans have access to the opportunities broadband can provide."[36] Issues include reforming the Universal Service Fund, intercarrier compensation, federal assistance for broadband in Tribal lands, expanding existing broadband grant and loan programs at the Rural Utilities Service, enabling greater broadband connectivity in anchor institutions, and improved broadband adoption and utilization especially among disadvantaged and vulnerable populations.
- *Part III (National Purposes),* which "makes recommendations to maximize the use of broadband to address national priorities. This includes reforming laws, policies and incentives to maximize the benefits of broadband in areas where government plays a significant role."[37] National purposes include health care, education, energy and the environment, government performance, civic engagement, and public safety. Issues include telehealth and health IT, online learning and modernizing educational broadband infrastructure, digital literacy and job training, smart grid and smart buildings, federal support for broadband in small businesses, telework within the federal government, cybersecurity and protection of critical broadband infrastructure, copyright of public digital media, interoperable public safety communications, next generation 911 networks, and emergency alert systems.

The release of the National Broadband Plan is seen by many as a precursor towards the development of a national broadband policy—whether comprehensive or piecemeal—that will likely be shaped and developed by Congress, the FCC, and the Administration.[38] Congress will likely play a major role in implementing the National Broadband Plan, both by considering legislation to implement NBP recommendations, and by overseeing broadband activities conducted by the FCC and executive branch agencies.

CURRENT FEDERAL BROADBAND PROGRAMS

With the conclusion of grant and loan awards established by the American Recovery and Reinvestment Act of 2009 (P.L. 111-5),[39] there remain two ongoing federal vehicles which direct federal money to fund broadband infrastructure: the broadband and telecommunications programs at the Rural Utilities Service (RUS) of the U.S. Department of Agriculture and the Universal Service Fund (USF) programs under the Federal Communications Commission (FCC). Although the USF's High Cost Program does not explicitly fund broadband infrastructure, subsidies are used, in many cases, to upgrade existing telephone networks so that they are capable of delivering high-speed services. Additionally, subsidies provided by USF's Schools and Libraries Program and Rural Health Care Program are used for a variety of telecommunications services, including broadband access. Currently the USF is undergoing a major transition to the Connect America Fund, which is targeted to the deployment, adoption, and use of both fixed and mobile broadband.

Table 4 (at the end of this report) shows selected federal domestic assistance programs throughout the federal government that currently can be associated with broadband and telecommunications development. The table categorizes the programs in three ways: programs exclusively devoted to the deployment of broadband infrastructure; programs which have traditionally focused on deployment of telecommunications infrastructure generally (which typically can and does include broadband); and applications-specific programs which fund some aspect of broadband access or adoption as a means towards supporting a particular application, such as distance learning or telemedicine.

Rural Utilities Service Programs

RUS implements two programs specifically targeted at providing assistance for broadband infrastructure deployment in rural areas: the Rural Broadband Access Loan and Loan Guarantee Program and Community Connect Broadband Grants.[40] The 110th Congress reauthorized and reformed the Rural Broadband Access Loan and Loan Guarantee program as part of the 2008 farm bill (P.L. 110-234). The 112th Congress considered reauthorization of the program as part of the 2012 farm bill.[41]

RUS also has a rural telephone loan program (dating back to 1949, now called Telecommunications Infrastructure Loans) that has historically supported infrastructure for telephone voice service, but has now evolved into support for broadband-capable service provided by traditional telephone borrowers. Additionally, the Distance Learning and Telemedicine Grant Program supports broadband-based applications.[42]

The Universal Service Concept and the FCC[43]

Since its creation in 1934 the Federal Communications Commission (FCC) has been tasked with "mak[ing] available, so far as possible, to all the people of the United States ... a rapid, efficient, Nation-wide, and world-wide wire and radio communications service with adequate facilities at reasonable charges."[44] This mandate led to the development of what has come to be known as the universal service concept.

The universal service concept, as originally designed, called for the establishment of policies to ensure that telecommunications services are available to all Americans, including those in rural, insular and high cost areas, by ensuring that rates remain affordable. Over the years this concept fostered the development of various FCC policies and programs to meet this goal. The FCC offers universal service support through a number of direct mechanisms that target both providers of and subscribers to telecommunications and, more recently, broadband services.[45]

Universal Service and the Telecommunications Act of 1996

Passage of the Telecommunications Act of 1996 (P.L. 104-104) codified the long-standing commitment by U.S. policymakers to ensure universal service in the provision of telecommunications services.

The Schools and Libraries, and Rural Health Care Programs

Congress, through the 1996 act, not only codified, but also expanded the concept of universal service to include, among other principles, that elementary and secondary schools and classrooms, libraries, and rural health care providers have access to telecommunications services for specific purposes at discounted rates. (See §§254(b)(6) and 254(h)of the 1996 Telecommunications Act, 47 U.S.C. 254.)

1. The Schools and Libraries Program. Under universal service provisions contained in the 1996 act, elementary and secondary schools and classrooms and libraries are designated as beneficiaries of universal service discounts. Universal service principles detailed in Section 254(b)(6) state that

"Elementary and secondary schools and classrooms ... and libraries should have access to advanced telecommunications services." The act further requires in Section 254(h)(1)(B) that services within the definition of universal service be provided to elementary and secondary schools and libraries for education purposes at discounts, that is at "rates less than the amounts charged for similar services to other parties."

The FCC established the Schools and Libraries Division within the Universal Service Administrative Company (USAC) to administer the schools and libraries or "E (education)-rate" program to comply with these provisions. Under this program, eligible schools and libraries receive discounts ranging from 20% to 90% for telecommunications services depending on the poverty level of the school's (or school district's) population and its location in a high cost telecommunications area. The FCC established a funding ceiling, or cap, of $2.25 billion, adjusted for inflation prospectively beginning with funding year 2010. Three categories of services are eligible for discounts: internal connections (e.g., wiring, routers and servers); Internet access; and telecommunications and dedicated services, with the third category receiving funding priority. According to data released by program administrators, approximately $31 billion in funding has been committed over the first 14 years of the program with funding released to all states, the District of Columbia and all territories. Funding commitments for funding Year 2012 (July 1, 2012-June 30, 2013), the 15th year of the program, totaled $ 2.6 billion as of July 5, 2013.[46]

2. The Rural Health Care Programs. Section 254(h) of the 1996 act requires that public and nonprofit rural health care providers have access to telecommunications services necessary for the provision of health care services at rates comparable to those paid for similar services in urban areas. Subsection 254(h)(1) further specifies that "to the extent technically feasible and economically reasonable" health care providers should have access to advanced telecommunications and information services. The FCC established the Rural Health Care Division (RHCD) within the USAC to administer the universal support program to comply with these provisions. Under FCC established rules only public or non-profit health care providers are eligible to receive funding. Eligible health care providers, with the exception of those requesting only access to the Internet, must also be located in a rural area. The funding ceiling, or cap, for this support was established at $400 million annually. The funding level for Year One of the program (January 1998-June 30, 1999) was set at $100 million. Due to less than anticipated demand, the FCC established a $12 million funding level for the second year (July 1, 1999

to June 30, 2000) of the program but has since returned to a $400 million yearly cap. As of March 31, 2012, a total of $514.3 million has been committed since the program's inception in 1998. The primary use of the funding is to provide reduced rates for telecommunications and information services necessary for the provision of health care.[47] In addition, the FCC established, in 2007, the "Rural Health Care Pilot Program" to help public and non-profit health care providers build state and region-wide broadband networks dedicated to the provision of health care services. There are 50 projects in the program with $387.9 million in authorized funds. As of February 29, 2010, $232.6 million of the funds have been committed to the 50 FCC designated projects. The FCC in a December 12, 2012, order, created a new program, the Healthcare Connect Fund, which will expand health care provider access to broadband, particularly in rural areas, and replace the Rural Health Care Pilot Program with a permanent program. Non-rural participation is limited to consortia of which at least 50% must be located in a rural area. The upfront funding cap for the Healthcare Connect Fund is $150 million annually and participants will be required to contribute 35% of the costs. The FCC also established, as part of the Healthcare Connect Fund, a new pilot program, to expand broadband connections to skilled nursing facilities. Funding for this pilot program, which will begin in 2014, will be up to $50 million total over three years. The total funding cap for all of the above mentioned USF rural health care programs will remain at $400 million annually.[48]

Universal Service and Broadband

One of the policy debates surrounding universal service in the last decade was whether access to advanced telecommunications services (i.e., broadband) should be incorporated into universal service objectives. The term universal service, when applied to telecommunications, refers to the ability to make available a basket of telecommunications services to the public, across the nation, at a reasonable price.

As directed in the 1996 Telecommunications Act (§254[c]) a federal-state Joint Board was tasked with defining the services which should be included in the basket of services to be eligible for federal universal service support; in effect using and defining the term "universal service" for the first time. The Joint Board's recommendation, which was subsequently adopted by the FCC in May 1997, included the following in its universal service package: voice grade access to and some usage of the public switched network; single line service; dual tone signaling; access to directory assistance; emergency service such as 911; operator services; and access and interexchange (long distance) service.

Some policy makers expressed concern that the FCC-adopted definition is too limited and does not take into consideration the importance and growing acceptance of advanced services such as broadband and Internet access. They point to a number of provisions contained in the Universal Service section of the 1996 act to support their claim. Universal service principles contained in Section 254(b)(2) state that "Access to advanced telecommunications services should be provided to all regions of the Nation." The subsequent principle (b)(3) calls for consumers in all regions of the nation including "low-income" and those in "rural, insular, and high cost areas" to have access to telecommunications and information services including "advanced services" at a comparable level and a comparable rate charged for similar services in urban areas. Such provisions, they state, dictate that the FCC expand its universal service definition.

The 1996 act does take into consideration the changing nature of the telecommunications sector and allows for the universal service definition to be modified if future conditions warrant. Section 254(c)of the act states that "universal service is an evolving level of telecommunications services" and the FCC is tasked with "periodically" reevaluating this definition "taking into account advances in telecommunications and information technologies and services." Furthermore, the Joint Board is given specific authority to recommend "from time to time" to the FCC modification in the definition of the services to be included for federal universal service support. The Joint Board, on November 19, 2007, concluded such an inquiry and recommended that the FCC change the mix of services eligible for universal service support. The Joint Board recommended, among other things, that "the universal availability of broadband Internet services" be included in the nation's communications goals and hence be supported by federal universal service funds.[49]

This debate was put to rest when provisions contained in the American Recovery and Reinvestment Act of 2009 (ARRA) called for the FCC to develop, and submit to Congress, a national broadband plan to ensure that every American has "access to broadband capability."[50] The FCC in its national broadband plan, *Connecting America: the National Broadband Plan,* recommended that access to and adoption of broadband be a national goal. Furthermore the national broadband plan proposed that the Universal Service Fund be restructured to become a vehicle to help reach this goal. The FCC, in an October 2011 decision, adopted an Order that calls for the USF to be transformed, in stages, over a multi-year period, from a mechanism to support voice telephone service to one that supports the deployment, adoption, and utilization of both fixed and mobile broadband.

This transformation includes the phase out of the USF's legacy High Cost Program and the creation of a new fund, the Connect America Fund, to replace it.[51]

LEGISLATION IN THE 110TH CONGRESS

In the 110th Congress, legislation was enacted to provide financial assistance for broadband deployment. Of particular note is the reauthorization of the Rural Utilities Service (RUS) broadband loan program, which was enacted as part of the 2008 farm bill (P.L. 110-234). In addition to reauthorizing and reforming the RUS broadband loan program, P.L. 110-234 contains provisions establishing a National Center for Rural Telecommunications Assessment and requiring the FCC and RUS to formulate a comprehensive rural broadband strategy.

The Broadband Data Improvement Act (P.L. 110-385) was enacted by the 110th Congress and required the FCC to collect demographic information on unserved areas, data comparing broadband service with 75 communities in at least 25 nations abroad, and data on consumer use of broadband. The act also directed the Census Bureau to collect broadband data, the Government Accountability Office to study broadband data metrics and standards, and the Department of Commerce to provide grants supporting state broadband initiatives.

Meanwhile, the America COMPETES Act (H.R. 2272) was enacted (P.L. 110-69) and contained a provision authorizing the National Science Foundation (NSF) to provide grants for basic research in advanced information and communications technologies. Areas of research included affordable broadband access, including wireless technologies. P.L. 110-69 also directs NSF to develop a plan that describes the current status of broadband access for scientific research purposes.

LEGISLATION IN THE 111TH CONGRESS

In the 111th Congress, legislation was enacted that sought to provide financial assistance for broadband deployment. Of particular note, provisions in the American Recovery and Reinvestment Act of 2009 (P.L. 111-5) provided grants and loans to support broadband access and adoption in unserved and underserved areas.

P.L. 111-5: The American Recovery and Reinvestment Act of 2009

On February 17, 2009, President Obama signed P.L. 111-5, the American Recovery and Reinvestment Act (ARRA). Broadband provisions of the ARRA provided a total of *$7.2 billion*, for broadband grants, loans, and loan/grant combinations.

The total consisted of $4.7 billion to NTIA/DOC for a newly established Broadband Technology Opportunities Program (grants) and $2.5 billion to the RUS/USDA Broadband Initiatives Program (grants, loans, and grant/loan combinations).[52]

Regarding the $2.5 billion to RUS/USDA broadband programs, the ARRA specified that at least 75% of the area to be served by a project receiving funds shall be in a rural area without sufficient access to high speed broadband service to facilitate economic development, as determined by the Secretary of Agriculture.

Priority was given to projects that provide service to the most rural residents that do not have access to broadband services. Priority was also given to borrowers and former borrowers of rural telephone loans.

Of the $4.7 billion appropriated to NTIA:

- $4.35 billion was directed to a competitive broadband grant program, of which not less than $200 million shall be available for competitive grants for expanding public computer center capacity (including at community colleges and public libraries); not less than $250 million to encourage sustainable adoption of broadband service; and $10 million transferred to the Department of Commerce Office of Inspector General for audits and oversight; and

- $350 million was directed for funding the Broadband Data Improvement Act (P.L. 110-385) and for the purpose of developing and maintaining a broadband inventory map, which shall be made accessible to the public no later than two years after enactment. Funds deemed necessary and appropriate by the Secretary of Commerce may be transferred to the FCC for the purposes of developing a national broadband plan, which shall be completed one year after enactment.

Final BTOP and BIP program awards were announced by September 30, 2010.

Other Enacted Broadband Legislation in the 111th Congress

P.L. 111-8 (H.R. 1105). Omnibus Appropriations Act, 2009. Appropriates to RUS/USDA $15.619 million to support a loan level of $400.487 million for the Rural Broadband Access Loan and Loan Guarantee Program, and $13.406 million for the Community Connect Grant Program.

To the FCC, designates not less than $3 million to establish and administer a State Broadband Data and Development matching grants program for state-level broadband demand aggregation activities and creation of geographic inventory maps of broadband service to identify gaps in service and provide a baseline assessment of statewide broadband deployment. Passed House February 25, 2009. Passed Senate March 10, 2009. Signed by President, March 12, 2009.

P.L. 111-32 (H.R. 2346). Supplemental Appropriations Act, 2009. Provides not less than $3 million to the FCC to develop a national broadband plan pursuant to the American Recovery and Reinvestment Act of 2009. Introduced May 12, 2009; referred to Committee on Appropriations. Passed House May 14, 2009; passed Senate May 21, 2009. Signed by President, June 24, 2009.

P.L. 111-80 (H.R. 2997). Agriculture, Rural Development, Food and Drug Administration, and Related Agencies Appropriations Act, 2010. For Rural Utilities Service, U.S. Department of Agriculture, provides $28.96 million to support a loan level of $400 million for the broadband loan program, and $17.97 million for broadband community connect grants. Introduced June 23, 2009; referred to Committee on Appropriations. Reported by Committee on Appropriations June 23, 2009. Passed House July 9, 2009. Passed Senate

August 4, 2009. Conference Report (H.Rept. 111-279) printed September 30, 2009. Signed by President October 21, 2009.

LEGISLATION IN THE 112TH CONGRESS

The 112th Congress examined the efficacy of federal broadband assistance programs and how they may fit into the context of a national broadband policy.

P.L. 112-10 (H.R. 1473). Department of Defense and Full-Year Continuing Appropriations Act, 2011. Rescinds existing unobligated past-year funding for the Rural Broadband Access Loan and Loan Guarantee Program and the Community Connect Grants at the Rural Utilities Service. For FY2011, appropriates $22.3 million to the Rural Broadband Access Loan and Loan Guarantee Program for the cost of broadband loans, and $13.4 million to Community Connect Grants. Signed by President, April 15, 2011.

P.L. 112-55 (H.R. 2112). Consolidated and Further Continuing Appropriations Act, 2012. Provides FY2012 appropriations for Rural Utilities Service broadband loan program and broadband community connect grants: $6 million for the broadband loan program (subsidizing a loan level of $212 million) and $10.372 million for Community Connect grants. Introduced June 3, 2011; referred to Committee on Appropriations. Reported by Committee on Appropriations June 3, 2011 (H.Rept. 112-101). Passed House June 16, 2011. Reported by Senate Appropriations Committee September 7, 2011 (S.Rept. 112-73). Signed by President, November 18, 2011.

H.R. 1083 (Owens). Rural Broadband Initiative Act. Establishes an Office of Rural Broadband Initiatives in the Department of Agriculture which would administer the RUS broadband loan and grant programs, and would develop a comprehensive rural broadband strategy. Introduced March 15, 2011; referred to Committee on Agriculture and in addition to the Committee on Energy and Commerce.

H.R. 1343 (Bass). To return unused or reclaimed funds made available for broadband awards in the American Recovery and Reinvestment Act of 2009 to the Treasury of the United States. Introduced April 4, 2011; referred to Committee on Energy and Commerce and to Committee on Agriculture. Reported (amended) by the Committee on Energy and Commerce (H.Rept. 112-228) on September 29, 2011. Passed House October 5, 2011. Referred to Senate Committee on Commerce, Science and Transportation October 6, 2011.

H.R. 1695 (Eshoo). Broadband Conduit Deployment Act of 2011. Directs the Secretary of Transportation to require that broadband conduit be installed as part of certain highway construction projects. Introduced May 3, 2011; referred to Committee on Transportation and Infrastructure.

H.R. 2163 (Matsui). Broadband Affordability Act of 2011. Amends the Communications Act of 1934 to establish a Lifeline Assistance Program for universal broadband adoption. Introduced June 14, 2011; referred to Committee on Energy and Commerce.

H.R. 5973 (Kingston). Agriculture, Rural Development, Food and Drug Administration, and Related Agencies Appropriations Act, 2013. For Rural Utilities Service, U.S. Department of Agriculture, provides $2 million to support a loan level of $21 million for the broadband loan program, and $10 million for broadband community connect grants. Introduced June 20, 2012; referred to Committee on Appropriations. Reported by Committee on Appropriations June 20, 2012.

H.R. 6083 (Lucas). Federal Agriculture Reform and Risk Management Act of 2012. Reauthorizes rural broadband loan program at $25 million per year through FY2017. Introduced July 9, 2012; referred to Committee on Agriculture. Ordered to be reported by committee July 11, 2012.

S. 257 (Landrieu). Small Business Broadband and Emerging Information Technology Enhancement Act of 2011. Seeks to improve certain programs of the Small Business Administration to better assist small business customers in accessing broadband technology. Introduced February 2, 2011; referred to Committee on Small Business and Entrepreneurship.

S. 1659 (Ayotte). To return unused or reclaimed funds made available for broadband awards in the American Recovery and Reinvestment Act of 2009 to the Treasury of the United States. Introduced October 5, 2011; referred to Committee on Commerce, Science and Transportation.

S. 1939 (Klobuchar). Broadband Conduit Deployment Act of 2011. Directs the Secretary of Transportation to require that broadband conduit be installed as part of certain highway construction projects. Introduced December 1, 2011; referred to Committee on Environment and Public Works.

S. 2298 (Brown of Ohio). Connecting Rural America Act. Amends the Rural Electrification Act of 1936 to improve the program of access to broadband telecommunications services in rural areas. Introduced April 18, 2002; referred to Committee on Agriculture, Nutrition, and Forestry.

S. 2375 (Kohl). Agriculture, Rural Development, Food and Drug Administration, and Related Agencies Appropriations Act, 2013. For Rural Utilities Service, U.S. Department of Agriculture, provides $6 million to support a loan level of $63 million for the broadband loan program, and $10 million for broadband community connect grants. Introduced April 26, 2012; referred to Committee on Appropriations. Reported by Committee on Appropriations April 26, 2012.

S. 3240 (Stabenow). Agriculture Reform, Food, and Jobs Act of 2012. Authorizes broadband loan and grant program at $50 million per year through FY2017. Introduced May 24, 2012; referred to Committee on Agriculture, Nutrition and Forestry. Reported to Senate May 24, 2012. Passed Senate (amended) June 21, 2012.

S. 3439 (Snowe). Federal Wi-Net Act. Directs the Administrator of General Services to install Wi-Fi hotspots and wireless neutral host systems in all federal buildings in order to improve in-building wireless communications coverage and commercial network capacity by offloading wireless traffic onto wireline broadband networks. Introduced July 25, 2012; referred to Committee on Environment and Public Works.

LEGISLATION IN THE 113TH CONGRESS

The following is a listing of broadband legislation directly related to the issue of federal assistance for broadband deployment in unserved areas.[53]

P.L. 113-6 (H.R. 933). Consolidated and Further Continuing Appropriations Act, 2013.Funds the broadband loan program at $4 million (supporting a loan level of approximately $42 million) and the Community Connect grant program at $10.372 million. Signed by President March 26, 2013.

H.R. 2163 (Matsui). Broadband Adoption Act of 2013. Amends the Communications Act of 1934 to reform and modernize the Universal Service Fund Lifeline Assistance Program. Introduced April 26, 2013; referred to Committee on Energy and Commerce.

H.R. 1947 (Lucas). Federal Agriculture Reform and Risk Management Act of 2013. Section 6105 would reauthorize the broadband loan and loan guarantee program through FY2018 at the current level of $25 million per year. Introduced May 13, 2013; reported by Committees on Agriculture and Judiciary.

H.R. 1639 (Gibson). Amends the Rural Electrification Act of 1936 to authorize loan/grant combinations under RUS broadband program. Introduced April 18, 2013; referred to Committee on Agriculture and Committee on Energy and Commerce.

H.R. 2410 (Aderholt). Agriculture, Rural Development, Food and Drug Administration, and Related Agencies Appropriations Act, 2014. Provides \$5.5 million to subsidize a loan level of \$42.146 million for the broadband loan program, and \$10.111 million for the Community Connect grant program. Introduced June 18, 2013; reported by House Committee on Appropriations (H.Rept. 113-116).

S. 954 (Stabenow). Agriculture Reform, Food, and Jobs Act of 2013. Includes the establishment of a new grant program in combination with the existing loan and loan guarantee program authorization, which is extended at \$50 million per year through FY2018. Passed by Senate June 10, 2013.

CONCLUSION

To the extent that the 113th Congress may consider various options for encouraging broadband deployment and adoption, a key issue is how to strike a balance between providing federal assistance for unserved and underserved areas where the private sector may not be providing acceptable levels of broadband service, while at the same time minimizing any deleterious effects that government intervention in the marketplace may have on competition and private sector investment.

In addition to loans, loan guarantees, and grants for broadband infrastructure deployment, a wide array of policy instruments are available to policymakers, including universal service reform, tax incentives to encourage private sector deployment, broadband bonds, demand-side incentives (such as assistance to low income families for purchasing computers), regulatory and deregulatory measures, and spectrum policy to spur roll-out of wireless broadband services. In assessing federal incentives for broadband deployment, the 113th Congress may consider the appropriate mix of broadband deployment incentives to create jobs in the short and long term, the extent to which incentives should target next-generation broadband technologies, the extent to which "underserved" areas with existing broadband providers should receive federal assistance, and whether broadband stimulus projects are being efficiently managed and how they may fit into the context of overall goals for a national broadband policy.

Table 4. Selected Federal Domestic Assistance Programs Related to Broadband and Telecommunications Development

Program	Agency	Description	Funding Amount (est. FY2012 unless otherwise noted)	Web Links
		Broadband Infrastructure Deployment Programs		
Broadband Technology Opportunities Program (BTOP)	National Telecommunications and Information Administration, Dept. of Commerce	Provides competitive grants to public and private sector entities in order to provide broadband access in unserved and underserved areas; provide broadband support and services to strategic institutions; improve broadband access by public safety agencies; and stimulate broadband demand, economic growth, and job creation.	$4 billion (ARRA, P.L. 111-5) (2009)	http://www.ntia.doc.gov/ broadbandgrants/
Broadband Initiatives Program (BIP)	Rural Utilities Service, U.S. Dept. of Agriculture	Provides competitive grants, loans, and loan/grant combinations to public and private sector entities in order to provide broadband access in unserved and underserved rural areas.	$2.5 billion for the cost of loans, grants, and loan/grant combinations (ARRA, P.L. 111-5) (2009)	http://www.rurdev.usda.gov/utp_bip.html
Rural Broadband Access Loan and Loan Guarantee Program	Rural Utilities Service, U.S. Dept. of Agriculture	Provides loan and loan guarantees for facilities and equipment providing broadband service in rural communities	$169 million for cost of money loans	http://www.rurdev.usda.gov/utp_farmbill.html
Community Connect Broadband Grants	Rural Utilities Service, U.S. Dept. of Agriculture	Provides grants to applicants proposing to provide broadband service on a "community-oriented connectivity" basis to rural communities of under 20,000 inhabitants.	$10 million	http://www.rurdev.usda.gov/utp_commconnect.html

Table 4. (Continued)

Program	Agency	Description	Funding Amount (est. FY2012 unless otherwise noted)	Web Links
	Telecommunications Broadband Infrastructure Deployment Programs			
Telecommunications Infrastructure Loan Program	*Rural Utilities Service, U.S. Dept. of Agriculture*	Provides long-term direct and guaranteed loans to qualified organizations for the purpose of financing the improvement, expansion, construction, acquisition, and operation of telephone lines, facilities, or systems to furnish and improve telecommunications service in rural areas. All facilities financed must be capable of supporting broadband services.	$145 million for hardship loans; $250 million for cost of money loans; and $295 million for FFB Treasury loans	http://www.rurdev.usda.gov/ utp_infrastructure.html
Distance Learning and Telemedicine Loans and Grants	*Rural Utilities Service, U.S. Dept. of Agriculture*	Provides seed money to rural community facilities (e.g., schools, libraries, hospitals) for advanced telecommunications systems that can provide health care and educational benefits to rural areas	$21 million	http://www.rurdev.usda.gov/ UTP_DLT.html
Universal Service High Cost Program[a]	*Federal Communications Commission*	Provides funding to eligible telecommunications carriers to help pay for telecommunications services in high-cost, rural, and insular areas so that prices charged to customers are reasonably comparable across all regions of the nation.	$4.5 billion (annually through 2017)	http://www.usac.org/hc/
Universal Service Schools and Libraries	*Federal Communications*	Provides discounts for affordable telecommunications and Internet access	$2.25 billion (annually adjusted for inflation)	http://www.universalservice.org/sl/

Program	Agency	Description	Funding Amount (est. FY2012 unless otherwise noted)	Web Links
Program (i.e., E-rate)	*Commission*	services to ensure that schools and libraries have access to affordable telecommunications and information services.		
Universal Service Rural Health Care Pilot Program[b]	*Federal Communications Commission*	Provides funds to cover 85% of the cost of constructing statewide or regional broadband telehealth networks and of connecting those projects to dedicated nationwide broadband telehealth networks and the public Internet.	$418 million (annually)	http://www.usac.org/rhc-pilot-program/
Appalachian Area Development Program	*Appalachian Regional Commission*	Project grants to support self-sustaining economic development in the region's most distressed counties and areas. Includes funds for a Telecommunications Initiative involving projects that enable communities to capitalize on broadband access.	$56 million	http://www.arc.gov/index.do?nodeId= 21
States' Economic Development Assistance Program	*Delta Regional Authority*	Grants for self-sustaining economic development projects of eight states in Mississippi Delta region.	$11 million (2011)	http://grants.dra.gov/
Investments for Public Works and Economic Development Facilities	*Economic Development Administration, Dept. of Commerce*	Provides funding for construction of infrastructure in areas that are not attractive to private investment; most funding is for water and sewer infrastructure but some has been designated for telecommunications and broadband projects.	$112 million	http://www.eda.gov/

Table 4. (Continued)

Program	Agency	Description	Funding Amount (est. FY2012 unless otherwise noted)	Web Links
Library Services and Technology Act Grants to States	*Institute of Museum and Library Services, National Foundation on the Arts and the Humanities*	Provides funds for a wide range of library services including installation of fiber and wireless networks that provide access to library resources and services.	$156 million	http://www.imls.gov/programs/programs.shtm
Native American Library Services	*Institute of Museum and Library Services, National Foundation on the Arts and the Humanities*	Grants to support library services including electronically linking libraries to networks.	$4 million	http://www.imls.gov/applicants/grants/nativeAmerican.shtm
Choice Neighborhood Implementation Grants	*Office of the Assistant Secretary for Public and Indian Housing and Office of Multifamily Housing Programs, Dept. of Housing and Urban Development*	Helps communities transform neighborhoods by revitalizing severely distressed public and/or assisted housing. Grantees may use funds to provide unit-based broadband Internet connectivity.	$110 million	http://www.hud.gov/cn/
Special Education—	*Office of Special*	Supports development and application of	$30 million	http://www.ed.gov/abou

Program	Agency	Description	Funding Amount (est. FY2012 unless otherwise noted)	Web Links
Technology and Media Services for Individuals with Disabilities	*Education and Rehabilitative Services, Dept. of Education*	technology and education media activities for disabled children and adults		t/offices/list/osers/index.html?src=mr/
Telehealth Network Grants	*Health Resources and Services Administration, Department of Health and Human Services*	Grants to develop sustainable telehealth programs and networks in rural and frontier areas, and in medically unserved areas and populations.	$6 million	http://www.hrsa.gov/telehealth/
Telehealth Resource Center Grant Program	*Health Resources and Services Administration, Department of Health and Human Services*	Provides grants that support establishment and development of telehealth resource centers to assist health care providers in the development of telehealth services, including decisions regarding the purchase of advanced telecommunications services.	$4 million	http://www.hrsa.gov/telehealth/
Licensure Portability Grant Program	*Health Resources and Services Administration, Department of Health and Human Services*	Provides support for state professional licensing boards to develop and implement state policies that will reduce statutory and regulatory barriers to telemedicine.	$0.35 million	http://www.hrsa.gov/telehealth/
NLM Extramural Programs	*National Library of Medicine, National Institutes of Health,*	Provides funds to train professional personnel; strengthen library and information services; facilitate access to and delivery of	$62 million	http://www.nlm.nih.gov/ep/extramural.html

Table 4. (Continued)

Program	Agency	Description	Funding Amount (est. FY2012 unless otherwise noted)	Web Links
	Department of Health and Human Services	health science information; plan and develop advanced information networks; support certain kinds of biomedical publications; and conduct research in medical informatics and related sciences.		
National Environmental Information Exchange Network Grant Program	*Environmental Protection Agency*	Provides funding to states, territories, and federally recognized Indian Tribes to support the development of an Environmental Information Exchange Network, including broadband infrastructure.	$10 million	http://epa.gov/exchange network/grants/

Source: Compiled by CRS from FY2013 budget documents, the Catalog of Federal Domestic Assistance, and grants.

[a] The High Cost program will be phased out, and replaced in stages by the Connect America Fund. The program provides funding to eligible service providers to support the provision of affordable voice and broadband services, both fixed and mobile.

[b] The Rural Health Care program will be replaced by the Healthcare Connect Fund.

End Notes

[1] The term "digital divide" can also refer to international disparities in access to communications and information technology. This report focuses on domestic issues only.

[2] FCC, Internet Access Services: Status as of June 30, 2012, released May 2013, p. 17. Available at http://transition.fcc.gov/Daily_Releases/Daily_Business/2013/db0520/DOC-321076A1.pdf.

[3] Ibid.

[4] Initially, and for many years following, the FCC defined broadband (or more specifically "high-speed lines") as over 200 kilobits per second (kbps) in at least one direction, which was roughly four times the speed of conventional dialup Internet access. In recent years, the 200 kbps threshold was considered too low, and on March 19, 2008, the FCC adopted a report and order (FCC 08-89) establishing new categories of broadband speed tiers for data collection purposes. Specifically, 200 kbps to 768 kbps is considered "first generation," 768 kbps to 1.5 Mbps is "basic broadband tier 1," and increasingly higher speed tiers are broadband tiers 2 through 7 (tier seven is greater than or equal to 100 Mbps in any one direction). Tiers can change as technology advances.

[5] NTIA, U.S. Broadband Availability: June 2010 – June 2012, May 2013, p.5.available at http://www.ntia.doc.gov/files/ ntia/publications/usbb_avail_report_05102013.pdf.

[6] Federal Communications Commission, Eighth Broadband Progress Report, FCC 12-90, released August 21, 2012, p. 29, available at http://transition.fcc.gov/Daily_Releases/Daily_Business/2012/db0827/FCC-12-90A1.pdf.

[7] U.S. Department of Commerce, National Telecommunications and Information Administration, Blog, "Household Broadband Adoption Climbs to 72.4 Percent," June 6, 2013, available at http://www.ntia.doc.gov/blog/2013/householdbroadband-adoption-climbs-724-percent.

[8] Eighth Broadband Progress Report, p. 54.

[9] U.S. Department of Commerce, National Telecommunications and Information Administration, Exploring the Digital Nation: America's Emerging Online Experience, June 2013, p. 26, available at http://www.ntia.doc.gov/files/ntia/ publications/exploring_the_digital_nation_-_americas_emerging_online_experience.pdf.

[10] Smith, Aaron, Pew Internet & American Life Project, Home Broadband 2010, August 11, 2010, available at http://www.pewinternet.org/~/media//Files/Reports/2010/ Home%20broadband%202010.pdf.

[11] For more information on rural broadband and broadband programs at the Rural Utilities Service, see CRS Report RL33816, Broadband Loan and Grant Programs in the USDA's Rural Utilities Service, by Lennard G. Kruger.

[12] See for example: National Exchange Carrier Association (NECA), Trends 2006: Making Progress With Broadband, 2006, 26 p. Available at http://www.neca.org/media/trends_brochure_website.pdf.

[13] Federal Communications Commission, Eighth Broadband Progress Report, FCC 12-90, released August 21, 2012, p. 5, available at http://transition.fcc.gov/Daily_Releases/Daily_Business/2012/db0827/FCC-12-90A1.pdf.

[14] NTIA, National Broadband Map, Broadband Statistics Report: Broadband Availability in Urban vs. Rural Areas, January 2013, p. 7, available at http://www.broadbandmap.gov/download/ Broadband%20Availability%20in%20Rural%20vs%20Urban%20Areas.pdf. Also see NTIA, U.S. Broadband Availability: June 2010–June 2012, May 2013, p. 10-11, available at http://www.ntia.doc.gov/files/ntia/publications/ usbb_avail_report_ 05102013.pdf.

[15] U.S. Department of Commerce, National Telecommunications and Information Administration, Exploring the Digital Nation: America's Emerging Online Experience, June 2013, p. 26, available at http://www.ntia.doc.gov/files/ ntia/publications/exploring_ the_ digital_nation_-_americas_emerging_online_experience.pdf.

[16] The terrain of rural areas can also be a hindrance to broadband deployment because it is more expensive to deploy broadband technologies in a mountainous or heavily forested area. An additional added cost factor for remote areas can be the expense of "backhaul" (e.g., the "middle mile") which refers to the installation of a dedicated line which transmits a signal to and from an Internet backbone which is typically located in or near an urban area.

[17] Gillett, Sharon E., Massachusetts Institute of Technology, Measuring Broadband's Economic Impact, report prepared for the Economic Development Administration, U.S. Department of Commerce, February 28, 2006, p. 4.

[18] Crandall, Robert, William Lehr, and Robert Litan, The Effects of Broadband Deployment on Output and Employment: A Cross-sectional Analysis of U.S. Data, June 2007, 20 pp. Available at http://www3.brookings.edu/ views/papers/crandall/200706litan.pdf.

[19] Mark Dutz, Jonathan Orszag, and Robert Willig, The Substantial Consumer Benefits of Broadband Connectivity for U.S. Households, Internet Innovation Alliance, July 2009, p. 4, http://internetinnovation.org/files/special-reports/ CONSUMER_BENEFITS_OF_BROADBAND.pdf.

[20] OECD, OECD Broadband Portal. Available at http://www.oecd.org/sti/ict/broadband.

[21] OECD, Directorate for Science, Technology and Industry, The Development of Broadband Access in OECD Countries, October 29, 2001, 63 pp. For a comparison of government broadband policies, also see OECD, Directorate for Science, Technology and Industry, Broadband Infrastructure Deployment: The Role of Government Assistance, May 22, 2002, 42 pp.

[22] See Turner, Derek S., Free Press, Broadband Reality Check II: The Truth Behind America's Digital Divide, August 2006, pp. 8-11. Available at http://www.freepress.net/files/bbrc2-final.pdf; and Turner, Derek S., Free Press, 'Shooting the Messenger' Myth vs. Reality: U.S. Broadband Policy and International Broadband Rankings, July 2007, 25 pp., available at http://www.freepress.net/files/shooting_the_messenger.pdf.

[23] National Telecommunications and Information Administration, Fact Sheet: United States Maintains Information and Communication Technology (ICT) Leadership and Economic Strength, at http://www.ntia.doc.gov/ntiahome/press/ 2007/ICTleader_042407.html.

[24] See Wallsten, Scott, Progress and Freedom Foundation, Towards Effective U.S. Broadband Policies, May 2007, 19 pp. Available at http://www.pff.org/issues-pubs/pops/pop14.7 usbroadbandpolicy.pdf. Also see Ford, George, Phoenix Center, The Broadband Performance Index: What Really Drives Broadband Adoption Across the OECD?, Phoenix Center Policy Paper Number 33, May 2008, 27 pp.; available at http://www.phoenix-center.org/pcpp/PCPP33Final.pdf.

[25] See price and services and speed data on OECD Broadband Portal, available at http://www.oecd.org/sti/ict/broadband; see also Federal Communications Commission, Third Annual International Broadband Data Report, IB Docket No. 10-171, DA 12-1334, August 21, 2012, available at http://hraunfoss.fcc.gov/edocs_public/attachmatch/DA12-1334A1.pdf.

[26] Federal Communications Commission, Notice of Inquiry, "Concerning the Deployment of Advanced Telecommunications Capability to All Americans in a Reasonable and Timely Fashion, and possible Steps to Accelerate Such Deployment Pursuant to Section 706 of the Telecommunications Act of 1996," FCC 04-55, March 17, 2004, p. 6.

[27] FCC News Release, FCC Improves Data Collection to Monitor Nationwide Broadband Rollout, November 9, 2004. Available at http://hraunfoss.fcc.gov/edocs_public/attachmatch/DOC-254115A1.pdf.

[28] Federal Communications Commission, Notice Proposed Rulemaking, "Development of Nationwide Broadband Data to Evaluate Reasonable and Timely Deployment of Advanced Services to All Americans, Improvement of Wireless Broadband Subscribership Data, and Development of Data on Interconnected Voice Over Internet Protocol (VoIP) Subscribership," WC Docket No. 07-38, FCC 07-17, released April 16, 2007, 56 pp.

[29] FCC, News Release, "FCC Expands, Improves Broadband Data Collection," March 19, 2008. Available at http://hraunfoss.fcc.gov/edocs_public/attachmatch/DOC-280909A1.pdf.

[30] For more information on the national broadband mapping program and the State Broadband Data and Development Program, see http://www.ntia.doc.gov/broadband grants/broadbandmapping.html.

[31] Eighth Broadband Progress Report, p. 59-60.

[32] Ibid., p.171, 177.

[33] Available at http://www.broadband.gov/plan/For more information on the National Broadband Plan, see CRS Report R41324, The National Broadband Plan, by Lennard G. Kruger et al..

[34] Federal Communications Commission, Connecting America: The National Broadband Plan, March 17, 2010, p. 9.

[35] Ibid., p. 11.

[36] Ibid.

[37] Ibid.

[38] See for example, Office of Science and Technology Policy and National Economic Council, The White House, Four Years of Broadband Growth, June 2013, 26 pages, available at http://www.whitehouse.gov/sites/default/files/ broadband_report_final.pdf.

[39] See CRS Report R40436, Broadband Infrastructure Programs in the American Recovery and Reinvestment Act, by Lennard G. Kruger.

[40] For more information on these programs, see CRS Report RL33816, Broadband Loan and Grant Programs in the USDA's Rural Utilities Service, by Lennard G. Kruger.

[41] Ibid.

[42] See CRS Report R42524, Rural Broadband: The Roles of the Rural Utilities Service and the Universal Service Fund, by Angele A. Gilroy and Lennard G. Kruger.

[43] The section on universal service was prepared by Angele Gilroy, Specialist in Telecommunications, Resources, Science and Industry Division. For more information on universal service, see CRS Report RL33979, Universal Service Fund: Background and Options for Reform, by Angele A. Gilroy.

[44] Communications Act of 1934, As Amended, Title I §1 [47 U.S.C. 151].

[45] Many states participate in or have programs that mirror FCC universal service mechanisms to help promote universal service goals within their states.

[46] For additional information on this program, including funding commitments, see the E-rate website: http://www.universalservice.org/sl/.

[47] For additional information on this program, including funding commitments, see the RHCD website: http://www.universalservice.org/rhc/.

[48] For more details on the USF rural health care support mechanism and the newly established Healthcare Connect Fund see In the Matter of Rural Health Care Support Mechanism, WC Docket No. 02-60, Federal Communications Commission, adopted December 12, 2012. Available at http://hraunfoss.fcc.gov/edocs_public/attachmatch/FCC-12- 150A1.pdf.

[49] The Joint Board recommended that the definition of those services that qualify for universal service support be expanded and that the nation's communications goals include the universal availability of: mobility services (i.e., wireless voice); broadband Internet services; and voice services at affordable and comparable rates for all rural and non-rural areas. For a copy of this recommendation see http://hraunfoss.fcc.gov/edocs_public/attachmatch/FCC-07J4A1.pdf.

[50] American Recovery and Reinvestment Act of 2009, P.L. 111-5, sec. 6001 (k)(2)(D).

[51] For a detailed discussion of this Order and USF transition see CRS Report R42524, Rural Broadband: The Roles of the Rural Utilities Service and the Universal Service Fund, by Angele A. Gilroy and Lennard G. Kruger.

[52] For information on existing broadband programs at RUS, see CRS Report RL33816, Broadband Loan and Grant Programs in the USDA's Rural Utilities Service, by Lennard G. Kruger.

[53] For information on public safety wireless broadband legislation, see CRS Report R41842, Funding Emergency Communications: Technology and Policy Considerations, by Linda K. Moore.

INDEX

C

D

E

F

G

H

I

J

K

L

M

N

O

P

Q

R

S

T

U

V

W

Y